Francis George Heath

Where to find Ferns

With a special Chapter on the Ferns round London

Francis George Heath

Where to find Ferns
With a special Chapter on the Ferns round London

ISBN/EAN: 9783337186258

Printed in Europe, USA, Canada, Australia, Japan

Cover: Foto ©berggeist007 / pixelio.de

More available books at **www.hansebooks.com**

WHERE TO FIND FERNS

WITH A SPECIAL CHAPTER ON

THE FERNS ·ROUND LONDON.

BY

FRANCIS GEORGE HEATH,

EDITOR OF THE NEW EDITION OF GILPIN'S "FOREST SCENERY";

*Author of "The Fern Portfolio," "Autumnal Leaves," "Tree Gossip,"
"The Fern World," "My Garden Wild," "Our Woodland Trees,"
"Sylvan Spring," "The Fern Paradise," "Burnham Beeches,"
"Trees and Ferns," "Peasant Life," "The English Peasantry,"
&c. &c.*

ILLUSTRATED.

PUBLISHED UNDER THE DIRECTION OF
THE COMMITTEE OF GENERAL LITERATURE AND EDUCATION
APPOINTED BY THE SOCIETY FOR PROMOTING
CHRISTIAN KNOWLEDGE.

LONDON :
SOCIETY FOR PROMOTING CHRISTIAN KNOWLEDGE,
NORTHUMBERLAND AVENUE, CHARING CROSS, W.C. ;
43, QUEEN VICTORIA STREET, E.C. ;
26, ST. GEORGE'S PLACE, HYDE PARK CORNER, S.W.
BRIGHTON : 135, NORTH STREET.
NEW YORK: E. & J. B. YOUNG & CO.
1885.

CONTENTS.

---◆◇◆---

B 2

iv CONTENTS.

THE ILLUSTRATIONS.

ILLUSTRATIONS of all the species of British Ferns are included in this volume; and British Ferns, it must be remembered, include species which comprise a not inconsiderable portion of those to be found in many other parts of the world than the British Islands. The illustrations of these Ferns have been reduced from the outlines which form the basis of the coloured figures of "THE FERN PORTFOLIO," to which work this little pocket-book is intended to be a companion.

The high praise which the Press has bestowed upon both the design and execution of the *facsimile* illustrations of "THE FERN PORTFOLIO" may be allowed, the Author trusts, to bespeak commendation for the very carefully made reductions from those illustrations—photographically accurate and true in all but colour—included in "WHERE TO FIND FERNS."

The figures in this volume are a little less than one-third natural size.

To prove the accuracy of the photographic reductions of the figures, readers are invited to compare them, by the aid of a pocket magnifying-glass, with those of "THE FERN PORTFOLIO." The same process will enable the reader to discover any points of detail that, appearing in the accompanying descriptions in the text, may not readily be discerned by the unaided eye in the figures.

Illustrations are also given in the chapter entitled "Definitions of Terms," and will, it is believed, add force and clearness to the explanations of that chapter.

But a feature of this little volume, which the Author believes is quite new to Fern literature, is the illustration of the chapter on "Fern Habitats." That so especial a feature of "WHERE TO FIND FERNS" will be widely appreciated by lovers of the beautiful plants which form its subject, the Author confidently believes.

It is unusual for the author of a book to say anything about its price; but in this instance the price of the volume has been a careful subject of study, with the object of widely increasing the love for a recreation whose pursuit must exercise a wholesome and healthy influence upon the public mind—an influence which, at once purifying and elevating, is calculated to raise the thoughts to better things, leading the mind from a contemplation of the beauty of Nature to the great Giver of all good things.

The price, therefore, of the volume, bound in cloth, is fixed at EIGHTEENPENCE; and as in this respect it stands alone amongst books of its kind, only a very large sale can make its issue remunerative.

LONDON, *May*, 1885.

WHERE TO FIND FERNS.

I.—EXPLANATORY.

HE title of this little work will indicate its object. But some slight explanation of its especial aim is necessary. It is intended to be a pocket volume. It will not attempt to supersede larger and more detailed and descriptive fern-books. Yet, though it will assume on the part of its readers some general knowledge of the beautiful flowerless plants which form its subject, it will, for convenience-sake, give descriptive, or rather definitive, notes of the ferns whose habitats it will indicate.

Already, in such works as "THE FERN PORTFOLIO," "THE FERN PARADISE," and "THE FERN WORLD," the Author has given descriptive accounts, accompanied by coloured and other illustra-

tions, of all the species of British ferns ; and to go
over again the ground thus occupied—and occupied,
too, by other writers—would be unnecessary, and it
would be also impossible, obviously, to give either the
elaborate illustration or the information in those works
within the space of the present one, which is merely
intended to supply, within the narrowest possible limits,

indications of the habitats and of the distribution
throughout the country of our British ferns.

The Author is unaware of the existence of any similar
volume with just the aim of this one ; and hence its
raison-d'être.   Fern-hunting, to lovers of ferns, is one of
the most delightful of pastimes.   It gives zest to any
country walk, because it adds the attraction of a hobby
to the pleasure of being out of doors.   Life, in the
present age, is far too sedentary, and there exists too

great a tendency to sit in rooms with closed doors and windows. Some people seem almost to dread air in motion, and they become, in time, so little used to it that, at length, the body itself is brought into a morbid state, currents of air become "draughts," and cold and illness are the result. The air is the best friend we have, and in seeking outdoor pastimes in the country we obtain it in its best and purest form. The seeker after ferns must ride his hobby into the wildest and most out-of-the-way districts (page 2), and into the most delicious nooks of greenery—must climb hills, wind through valleys, plunge into woods, follow the course of streams, search rocks, hedgebanks, and forest-clumps, examine old walls and tree-forks, and look everywhere, in short, where green life has a chance of existence.

But many persons who have a general knowledge of ferns do not know the particular places in which the various species should be looked for ; and it would require the exercise of a very unusual memory to remember the particular districts over which the various species are distributed, or from which certain of the commoner kinds are excluded.

To supply such data in a concise form under the name of each fern, after first giving illustrated "Definitions of Terms," an illustrated chapter on "Fern Habitats," and a chapter on "The Cultivation of Ferns," is the especial purpose of this little volume. There will follow a special chapter on the "Ferns round London," and an alphabetical index of the particular localities mentioned in the other sections of the book.

II.—DEFINITIONS OF TERMS.

O render unnecessary the repetition of explanations of the meaning of the botanical terms used in the description of the parts of ferns, the definition of such terms as are used in this volume will be here given. The list will be as short as possible, because generally the simplest and least technical expressions will be used, and botanical terms will only be resorted to when they indicate what could only otherwise be conveyed by several words. By reading this short chapter once or twice the uninitiated will, therefore, be readily able to understand all that is said in the succeeding chapters, and will not find themselves involved—as they would were nothing but technical terms employed—in the mazes of a new language.

Ferns, as most of our readers will scarcely need to be reminded, are flowerless plants, allied to funguses, lichens, liverworts, mosses, and seaweeds, but standing higher than those orders in the scale of vegetation. Their more immediate allies are plants of the following orders :—*Equisetaceæ* (Horsetails) ; *Lycopodiaceæ* (Club-mosses) and *Marsileaceæ* (Pepperworts). All these plants belong to the large class designated, in the botanical arrangement of the vegetable kingdom, *Cryptogamia*—so designated because the fructification, produced without the agency of flowers, is more or less concealed by being borne on the backs or edges of their leafy parts.

Here is a portion—the under side of one of the *pinnæ*
(or branches) of the Broad Buckler Fern (*Lastrea dilatata*)
—showing conspicuously the fruit scattered like small
spots on its surface.   To see this fruit when present, the

fronds of a growing plant would have to be turned up
to the light.

Like other plants, ferns consist of three principal
parts—*roots*, *stems*, and *leafy parts*.   The accompanying

figure of the Scaly Spleenwort (*Asplenium ceterach*) illustrates the parts just mentioned. These, with their sub-

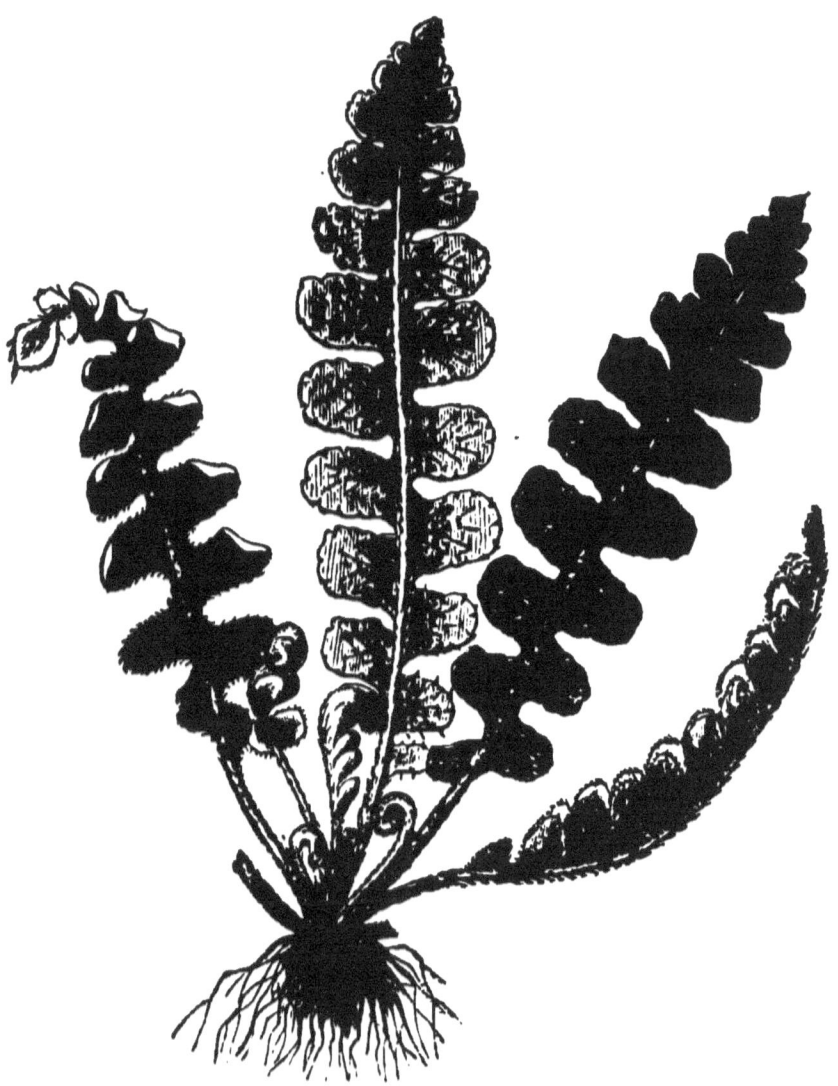

divisions and the organs or appendages connected with them, will be described as follows.

All ferns have *roots* which are more or less fibrous ;
being sometimes very fine, tough, and wiry, and sometimes thick, brittle, and fleshy.   The finer fibrous roots

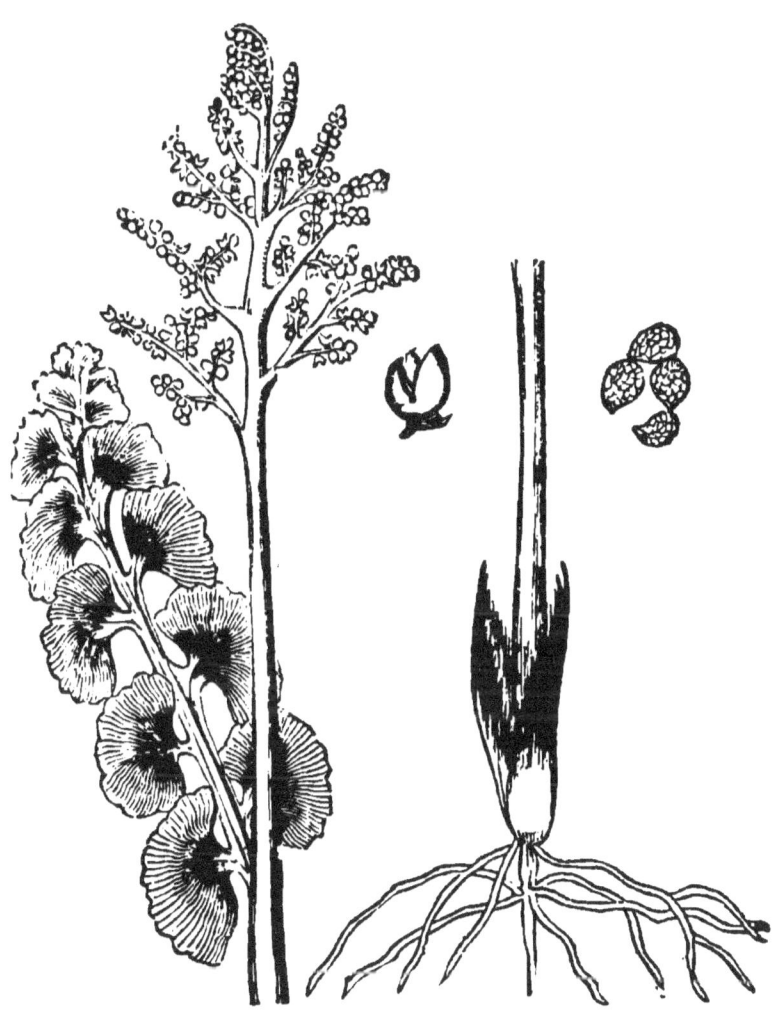

may be illustrated by those of the Scaly Spleenwort ;
the thick, brittle, and fleshy ones by those figured in the
sketch, on this page, of the Moonwort (*Botrychium*

lunaria). To get this figure, which is life-size, into our page, it is shown in two parts, the roots and part of the stem or *stipes* on the right-hand side, and the remainder of the stem (from the point of severance) and the barren and fertile fronds on the left-hand side.

The *stems* of ferns are of two principal kinds. The one kind is ordinarily called a *caudex* or *cormus*, the other a *rhizoma*. Strictly speaking, the *caudex* simply means the stem, of whatever kind. Many persons erroneously regard that part of a stem which is wholly or partially buried in the earth as a root. As even botanists give various and conflicting definitions of the parts of plants which are either roots or stems, it will be desirable, in this place, to make it clear in what sense the terms employed in the following chapters are used in relation to ferns.

When *roots* are referred to, it will be understood that the expression has reference, solely, to the fibrous underground parts of ferns, such as that shown in the figure of the Scaly Spleenwort (page 6).

The *rootstock* is the basal part of the *stem* from which, growing downwards, the roots spring. The upper part of the stem is called the *crown*. From this arise the leafy parts of ferns.

If the stem be more or less globular, bulb-shaped, and erect in habit, it is said to be a *cormus*. If it lies or creeps horizontally upon, or underneath, the soil, it is called a *rhizoma*.

The form and appearance of the *rhizoma* are shown in the figure, on page 9, of that very beautiful fern, the European Bristle Fern (*Trichomanes radicans*). In this figure the creeping stem is distinctly indicated, with its fibrous rootlets, together with one completely expanded, and three unrolling fronds underneath. The rhizoma, as the illustration also shows, is clothed with dark-coloured hair or down.

. Few of our native ferns have stems which rise more

than an inch or two above ground. When a stem rises to a height of many feet·above the ground it forms a *trunk*, becomes tree-like, and ferns of this habit are called tree-ferns. The elongation of a stem to form a trunk is a

process accomplished by the heightening of the crown of the cormus by the retention, each year, of the bases of the fronds which rise above it in a circlet. The older the fern, therefore, the higher, up to a certain limit, will

be the trunk; for, though the upper parts of the fronds die away, they leave the lower parts as contributions to the stem.

How beautiful are great tree-like forms of ferns (page 10) can only be fully appreciated by those who have seen these exquisitely-beautiful inhabitants of tropical forests in their native habitats.

The only British species that, in character, at all resembles a tree-fern is *Osmunda regalis*, which forms a trunk sometimes two feet in height.

From the upper parts of the stems of ferns rise the *fronds*, the name given to their leafy parts. The term *frond* will be here used to mean the leafy part and the long or short stalk which supports it and connects it with the crown. This stalk is called the *stipes*; but, when reference is made to the *shape* of the frond, it must be understood that *only* the leafy part is referred to.

In the ensuing illustration (page 12) of the Lady-Fern (*Athyrium filix-fœmina*) the leafy part is shown separately from the scaly stipes on the right-hand side.

The mid-stem of the frond, continuing from the stipes into the leafy part, is called the *rachis*. If this be branched, the principal or central mid-rib is the *primary rachis* and the branches are the *secondary rachides*.

If the frond assume the form of a single leaf with an unindented margin, it is said to be *simple*. The term *entire* is used to refer to an unindented margin.

In the figure, on page 13, of the Adders-tongue (*Ophioglossum vulgatum*) the oval leafy part illustrates what is called an *entire* margin.

When the frond is like a single leaf with incisions which, though deep, do not reach down to the rachis, it is described as being *pinnatifid*. Such is the form illustrated by the Scaly Spleenwort on page 6. If the indentations reach the rachis, leaving it bare, the frond becomes *pinnate*, and each separated leafy part is called

c

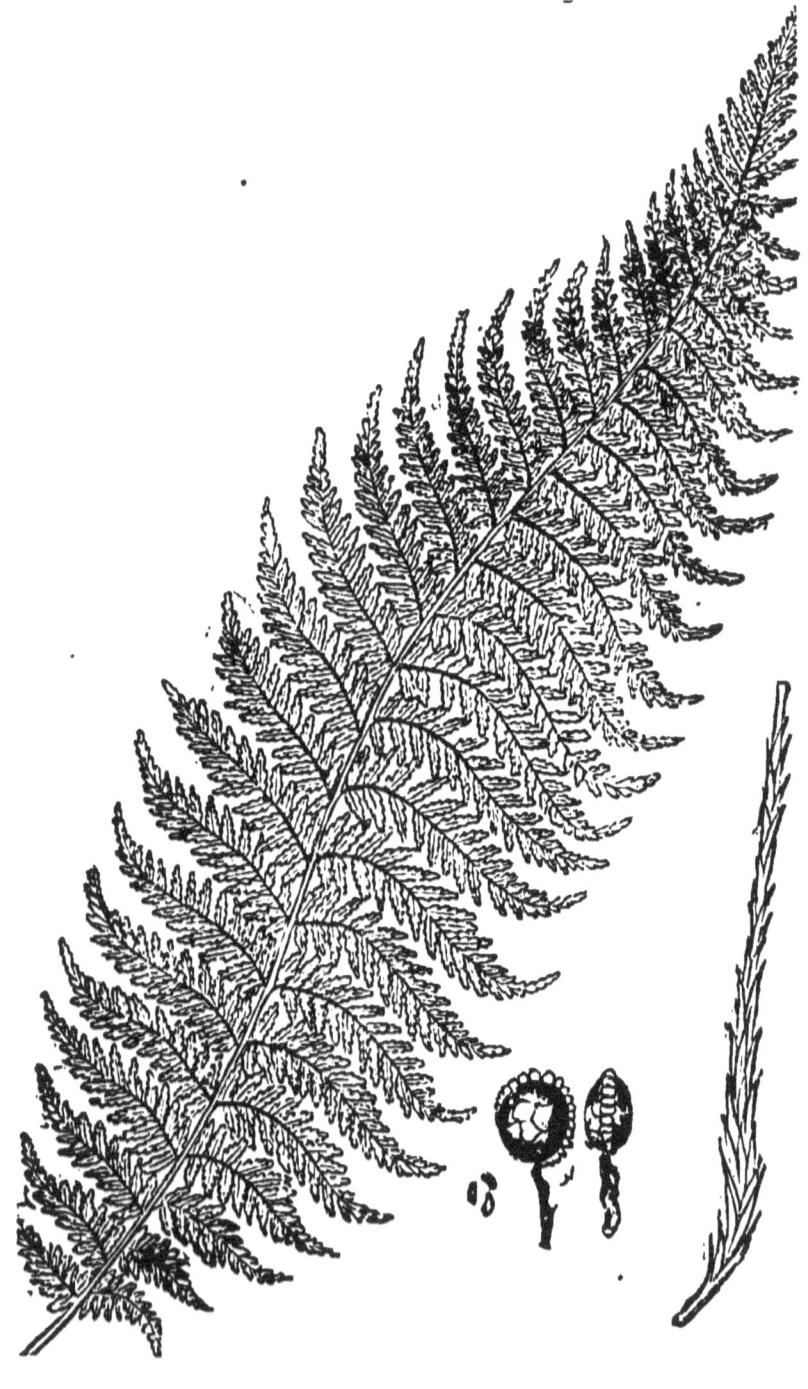

a *pinna*—all the parts thus separated being designated, in the aggregate, *pinnæ.* The pinnæ may be entire, simply or deeply indented, or again, and more elaborately, divided. If twice so divided, taking the entire form as the starting-point, the frond is *bi-pinnate,* and if once more, or thrice divided, it is *tri-pinnate.* It is *decompound* if more than three times divided. The parts into which pinnæ are immediately subdivided are termed *pinnules.* The immediate subdivisions of pinnules are *lobes.* Thus a thrice-divided frond, such as that of the Bracken, consists of stipes, rachis, secondary rachides, pinnæ, pinnules, and lobes.

The seeds of ferns, differing essentially from the seeds of flowering plants, are called *spores.* They are dust-like bodies infinitesimally small, and are enclosed—on the backs or along the under edges of the fronds in a particular order—in differently-shaped spore-cases called *sporangia.* The sporangia are generally produced in clusters or heaps called *sori,* each individual cluster being called a *sorus.*

In the figure of a pinna of the Broad Buckler Fern on page 5, the arrangement of these heaps or clusters of spore-cases was indicated. Here (page 14) is a magnified lobe of a pinnule of the same frond, magnified so as to show very clearly its form, and the form and position of each *sorus* with respect to the veins of the leafy part, and to show also the hairiness of the stem from which it springs and the incisions of the leafy

margin. The shape of the sorus differs in different ferns. It is mostly rounded; but in some ferns, as shown in the subjoined illustration of a portion of a frond of the Male

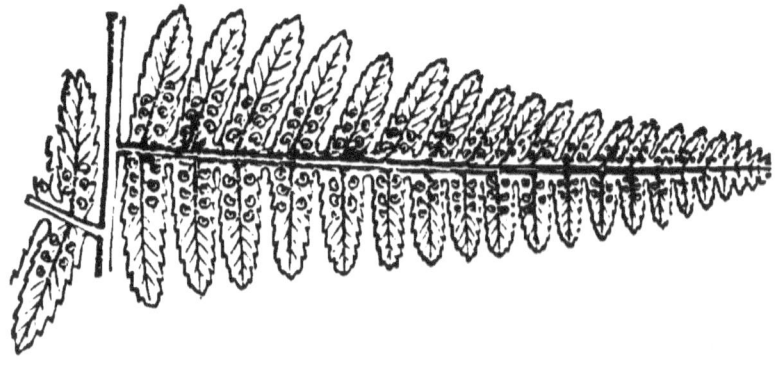

Fern (*Lastrea filix-mas*), it is kidney-shaped. The annexed figure, much enlarged from the natural size of a portion of a Male Fern frond, shows very clearly both the shape (and the position with regard to the veins) of the sorus.

Sometimes the sori are protected by scale-like coverings. Each such covering protecting a sorus is named an *indusium*. On page 15 is a drawing of an indusium of a sorus of the Broad Buckler Fern, one of the *Lastreas*. It is much magnified, but shows clearly the kidney shape and the jagged margin. Where there are no *indusia* the sori are said to be *non-indusiate*, or naked. In some species the margins of the pinnules are turned back over the sori and cover them after the manner of indusia. The fructification, in such cases, is produced close to the extreme outer edges of the leafy parts of the frond, and is then said to be *marginal.* In the case of those sori covered by  indusia when the spores are ripened, the indusia dry up and fall off, and the spore-cases enclosed themselves burst and liberate the infinitesimal germs they contain. We shall see presently what is the shape of some spores and spore-cases.

Fructification is a term applied to the general system of spores. Some fronds bear no fructification, in which case they are said to be *barren;* whilst others, upon the same plant, are spore-bearing, and these are called *fertile* fronds. The fructification, as we have seen by the magnified figures on page 14, is attached to the veins which ramify over the leafy substance of the frond. The system of veins is called the *venation.* That particular portion of the venation to which the fructification is attached bears the name of the *receptacle.*

Into a *detailed* consideration of the question of *classification* it is not the design of this volume to enter.

Ferns constitute a great *class* of the vegetable kingdom. According to one of our botanical systems this class is subdivided into *orders,* the orders into *genera,* the genera into *species,* the species into *varieties.* In the botanical arrangement of British plants under this particular system ferns belong to the third class—called *Acotyledons* or *Cryptogams* (the other two classes of plants being, 1, *Dicotyledons ;* and, 2, *Monocotyledons*). These collective expressions are used to indicate that the plants which are designated by them are produced from seeds which are of three kinds : 1, seeds which have each *two cotyledons*—a cotyledon being a seed-lobe, and, for assimilative purposes, a seed-leaf whether developed above ground or beneath the soil ; 2, seeds which have each *one* cotyledon; and, 3, seeds, *without* cotyledons, such as are the spores of cryptogamic plants.

Under the same system there are in the class of Acotyledons nine orders, of which ferns—*filices*—constitute the first. The orders are subdivided into *tribes*, the tribes into *genera*, and these into *species* and *varieties*. Under *filices* there are four tribes, nineteen genera, forty-five species, and almost endless varieties. Here we shall only take note of genera and species, and the descriptive and enumerative parts of the volume will thus be found divided into forty-five sections, headed by the common and botanical name of each species of British fern. But, before leaving the present chapter, we shall say a little about fern-spores, and indicate the classification adopted by botanists with regard to British ferns in so far as it is based upon the character of the spore-cases, and the character of the unrolling fronds.

British ferns, then, are divided for purposes of classification into three groups, named, 1, POLY-PODIACEÆ ; 2, OSMUNDACEÆ ; and, 3, OPHIOGLOSSACEÆ. The first group, *Polypodiaceæ*, includes ten smaller groups, comprising fifteen genera : viz. : *Polypodium*, *Allosurus*, *Gymnogramma*, *Polystichum*, *Lastrea*, *Athyrium*, *Asplenium*, *Scolopendrium*, *Blechnum*, *Pteris*, *Adiantum*, *Cystopteris*, *Woodsia*, *Trichomanes*, and *Hymenophyllum*. The spore-cases in this group are girt by an elastic ring which, on bursting, causes the spore-case to open by what is called a "transverse fracture." The form of the case, the elastic ring, the manner in which it opens, and the shape of the spores enclosed in it are illustrated by the diagrams which follow, and which exhibit the Common Polypody (*Polypodium vulgare*), with a portion of its rhizoma, a frond, a spore-case enormously enlarged, much enlarged spores, and enlarged pinnæ, exhibiting, in one case, the veins, and, in the other, enlarged, non-indusiate sori.

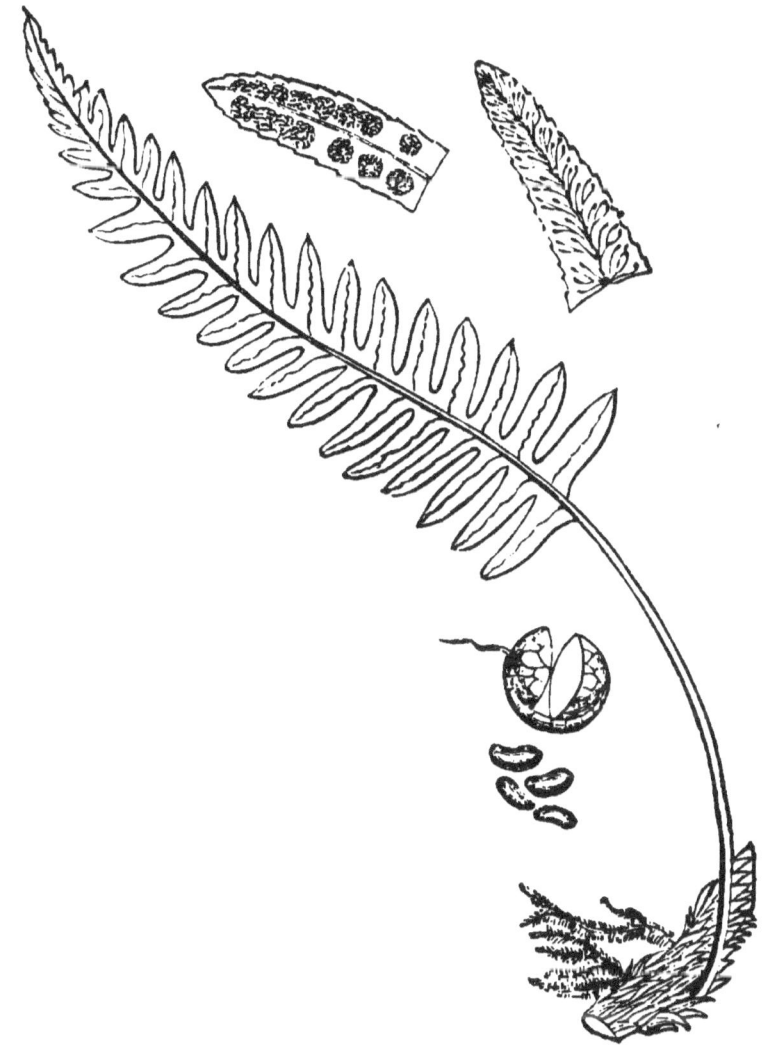

The manner of the fronds unrolling—a character which characterises the group—is circinate or scroll-like, and is shown in another genus belonging to the group, a species of which, *Asplenium ceterach*, has been previously figured on page 6.

The next group, *Osmundaceæ*, includes only one genus and one species in Britain, *Osmunda regalis;* and, though the fronds are rolled up in the same way as in *Polypodiaceæ*, there is no elastic ring around the spore-cases, and these are two-valved, and burst vertically.

The accompanying marginal cut will show the shape 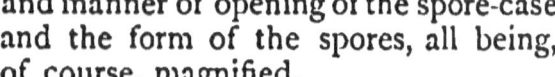 and manner of opening of the spore-case and the form of the spores, all being, of course, magnified.

In the third group, *Ophioglossaceæ*, are two genera and three species, viz.,

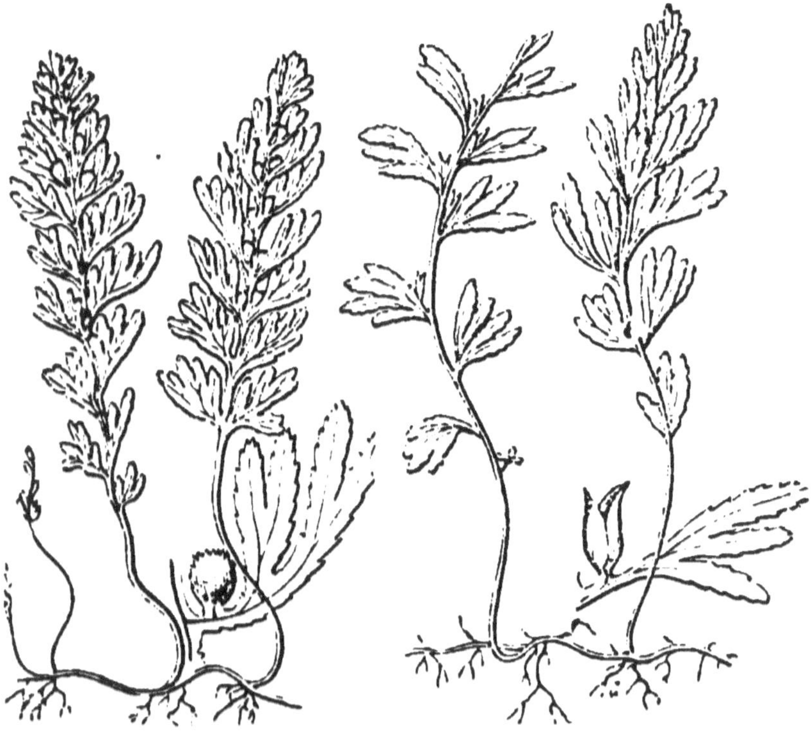

Botrychium lunaria and *Ophioglossum vulgatum* and *lusitanicum*. The fronds in this group are folded up straight, and the spore-cases are two-valved, and have no elastic ring, as will be seen on examining the figures given with the illustration, on page 7, of the Moonwort.

Accompanying the figure, given on page 9, of the European Bristle Fern, an enlarged diagram shows the urn-shaped and peculiar position of its receptacle. Through the centre of this receptacle the prolonged end of a vein passes, and on this vein are strung the spore-cases. The urn shape of the receptacle, in the case of the two Filmy Ferns, is also conspicuously shown. in the illustrations, on page 18, of those ferns.

III.—FERN HABITATS.

 HO can doubt that much of the fascinating attraction of the pursuit of ferns arises from the loveliness of the spots where they grow, and, to those new to the pastime, from the pleasurable surprise attendant upon finding forms. of beauty in places so dark and shadowy as to be half gloomy? Looking into such places,—hollows in rocks, openings in the leafy shrouds of hedge-banks, and the shadowy spaces which lie beneath the dense undergrowths of woods —the eye, at first, oftentimes sees nothing but the merest shadowy outlines. But, as it becomes accustomed to the darkness, it begins to discern the delicate, graceful, and feathery forms of some members of the great family of shade and moisture-loving plants. Looking still, the forms become·

bolder, until every curve and indentation stands out
with perfect distinctness. At other times the eye is
pleased with the wealth of beauty revealed to it by the
crowding of graceful ferny forms upon open hill-sides,
over sunlit forest glades, or upon the boulder-strewn
expanse of some rugged moorland. The country which
produces the most beautiful scenery furnishes in greatest
abundance the most lovely forms of fern life; and ferns
lend additional beauty to lovely scenery.

Yet ferns are often present in many places without
being seen. They are so modest and retiring in habit,
that they frequently hide, so to speak, in the most
sequestered nooks. But it is always easy to find them
when it is known where to look for them. Their powers
of reproduction are so great, the infinitesimal spores are
so easily wafted far and wide by the restless winds, when
the season of ripening has arrived and the bursting
sporangia have set at liberty the multitudes, infinitely
vast, of their imprisoned germs, that the presence even
of the rarest ferns is always *possible*, even in places least
suspected to possess them.

It may generally be assumed that, wherever ferns have
been once actually discovered, they will be found again,
if not in the immediate vicinity, at least somewhere in
the same neighbourhood. Even when well-known
habitats of rare ferns have been stripped of all promi-
nently visible specimens, the old ferns taken away are
almost certain to have had opportunities of shedding
their spores before their removal; and in a year or two,
when the minute seedlings have had time to assume
ferny forms, they may be looked for in the same spots
with a tolerable certainty of finding them, provided the
conditions of growth have not been changed by an
alteration in the character of the habitats.

With regard to several species of British ferns, the
recorded habitats are very few in number, and the species
in question are pronounced to be " rare." But, when it

is remembered that the opportunities of obtaining the topographical information which has been published in books on this subject have necessarily been limited, it may fairly be assumed that the habitats of these ferns are much more numerous than they are generally supposed to be. Small as are the British Islands, and thoroughly overrun as are most parts of them, there are, nevertheless, tens of thousands of places suitable for the growth of ferns that are practically *terra incognita*, though not by any means inaccessible to the fern-hunter. The Author of this volume has frequently, in the earlier days of his fern-hunting excursions, in looking for rare ferns in places to which experienced guides have directed him, by taking the trouble to look further in the same neighbourhood, come upon places surprisingly rich in specimens whose existence had been wholly unsuspected and obviously unknown. These " finds " have been due to careful notings of the favourite habitats of the species, and to the application of this knowledge to the practical working of a system of persistent and elaborate *search*. Yet the occupation has always been a pleasurable one, and has only been undertaken as a delightful holiday pastime.

What is true of small districts is likely to be equally true of large ones. The Author believes that many ferns, supposed to be entirely absent from certain parts of this country, are really present, but undiscovered. He has had many proofs, furnished to him by numerous correspondents, of the occurrence of certain ferns in counties and districts never before recorded as possessing them.

One especial feature of this volume will be its indications of the *particular positions* in which ferns grow, so that the *exact places* in which to look for the various species may be known. This information, derived from the Author's own knowledge, will be supplemented by the fullest possible lists of the counties—for the more widely distributed species—and of the smaller districts—for the rarer kinds—in which each fern has been discovered growing wild. For some of this information as to county localities, the Author is indebted to Mr. Hewett Watson's "Topographical Botany." To give minutely-detailed indications of the *exact spots* in which the various ferns

are to be found would be to destroy half the charm of fern-hunting.

The Author desires especially to impress upon those who may read and *use* this book that there is no outing in the country—however brief may be the period during which it lasts, and however apparently unpromising may be the district—that may not result in the finding of some ferns which may be none the less valued because they are *common*. The *rarity* of a "find" does, however, un-questionably give pleasure to the majority of fern-hunters. And such brief outings as have been referred to are sure to have great zest given to them by the possibility of finding a "prize" as the result of a minute and careful search in such places as those indicated in the im-mediately succeeding pages.

If, now, we can pictorially as well as verbally indicate the places in which the fern-lover may expect to find the object of his quest, we shall, we trust, impart a new pleasure to a delightful pursuit.

First, then, let us take the ever-abundant and delightful Bracken (*Pteris aquilina*) (page 22), which with feathery grace and beauty drapes wide areas of common, moor, and forest, fringing paths for miles in open glade and shady woodland path, as hardy and luxuriant as it is beautiful. It loves the sun as no fern does, and even in sunlit forest glades will sometimes rise so high on either side as to hide the tallest passer-by. It is by far the most abundant of all its kind, and is the most familiar to those who know least how to distinguish a fern from another plant.

On page 24 is a little peep of the Doone Glen, ren-dered immortal by Mr. Blackmore's fascinating story of "Lorna Doone." Upon just such upland slopes as those which rise from the stream that winds through this moor-land, the Bracken would be found, and down by the water's margin, in little stony but rich and moist nooks, one might look with confidence for the delightfully-scented golden green Mountain Buckler Fern (*Lastrea montana*). In similar nooks along the stream-bank, often growing in clumps with the Mountain Buckler Fern, would be also found the Hard Fern (*Blechnum spicant*). Under shelter of the trees, shown in the foreground of the picture, yet coying down as near as possible to the moor-land stream the Hartstongue (*Scolopendrium vulgare*) and the Lady Fern (*Athyrium filix-fœmina*) would be found growing singly or together in clumps, roots interlacing with roots. Under the same shady influence it must be strange if we did not come upon *Osmunda regalis*, but this would be in positions where the soil was more than usually peaty and soft, and where the rootlets could touch the percolating water.

Talking of Mr. Blackmore and his beautiful book tempts us to give a little glimpse, on page 25, of the Bagworthy water-slide at the foot of the same Doone Glen, a spot known by heart to thousands who have never seen the place, but whose recollections will never cease to vividly retain the graphic and awe-inspiring recital of Jan

Ridd's adventure up this famous slide to find love and Lorna in the terrible Doone Glen at its head. What lover of ferns could fail to recognise, in such a spot as this which we have just opened to view, a chosen abode of ferns? Here, at the foot of the rocks, the Lady Fern would revel in the moist and half-gloomy air. So would the

Hartstongue and the Hard Fern. Upon the rocks themselves we should find the common Maidenhair Spleenwort (*Asplenium trichomanes*), the Wall Rue (*Asplenium ruta-muraria*), the Common Polypody; and possibly, if careful search were made, the Mountain Polypody (*Polypodium phegopteris*) in the moister leaf-mould corners; the Black Maidenhair Spleenwort

(*Asplenium adiantum-nigrum*), too, nestling in sheltered
stony crevices. Then, up stream amongst the trees, not too
far up, but near the base of the fall, the Prickly-toothed
Buckler Fern (*Lastrea spinulosa*), and possibly a speci-

men or two of *Osmunda*. Many of these, especially the
rarer kinds, have perhaps been carried off by the
thousands of visitors to this 'enchanted spot. The
Author only knows that, when he visited the Doone Glen
and the water-slide, he saw many of the species he has

enumerated, and it is just such spots as these that should furnish the kinds of fern that have been named.

Talking of Devonshire, we must give one or two little pictures of its scenery in places certain to be crowded with many kinds of ferns.

Here (page 26) is a bit of the Plym, near Cadover Bridge. The river is brawling along just as Devonshire rivers like to brawl, softly and musically, though with great meaning, which implies the power to thunder when heavy rains upon the moors bring down the waters with a sudden rush that bears no resistance. Amongst the riverine boulders the fern-hunter will not look in vain probably, even if he have to search a little way, for *Osmunda* and *Blechnum*, Lady Fern and Male Fern (*Lastrea filix-mas*). Three other Buckler Ferns he is not unlikely to find,—the Mountain, the Prickly-toothed, and the Crested. The upland immediately beyond the water will certainly give him *Pteris aquilina*, and—not impossibly—careful search amongst the grassy roots would lead to the discovery of the Moonwort, and a little more towards the water, if he looks in somewhat moister positions than he expects to find *Botrychium lunaria* in, the Adders-tongue (*Ophioglossum vulgatum*).

Looking out now for wood and water, we could scarcely select a more typical bit of Devonshire fern country than the scene in Bickleigh Vale, represented on page 28. From these stony water-margins are sure to look out, their roots snugly embedded in the leaf-mould angles of their rocky habitats, grand specimens of Lady Fern, Hartstongue, Hard Fern, and, a little higher on the banks, the Hard and Soft Prickly Shield Ferns, the Common Polypody, the Broad Buckler Fern, and the Black Maidenhair Spleenwort. In this same neighbourhood, creeping along over moist stony surfaces, there should be found masses of the two Filmy Ferns, *Hymenophyllum tunbridgense* and *Hymenophyllum unilaterale*.

The beautiful Filmy Ferns, though absent from many

D

wide areas of country, are very abundant in those
spots in which the conditions of their delicate growth
are fulfilled. Amongst places known to the Author,
there are none where he has found them in such great
abundance as in the delicious bit of fern country lying
contiguous to Shaugh Bridge, that crosses the ferny Plym,

of which mention has just been made. Our illustration
(page 29) will give a glimpse of the boulder bed of
this pretty river. Not far from the bridge there rises
from the stream-level what may be termed a boulder
amphitheatre, consisting of great rocks, some smooth
and some rugged, and ranging in size from boulders like
those shown in our picture to giant rocks, that look as

if some giant hand had scattered the rocky hills around, and thrown the *débris* into the valley. Here and there tiny rills trickle down from the higher rocks to the river below, and in many a moist position, in rocky hollows

between rock and rock, and on the crest of the stony surfaces, the Filmy Ferns form dense carpetings. Veritable carpetings they are, for the fibrous roots and the extensively creeping rhizomas of the ferns are so thick and matted that they could be stripped from the rocks in sheets, though no fern-gatherer should take more than a modest share of what is intended for all.

Stony bridges no longer new, when the mortar begins to crumble, and leaf-mould to gather in the crevices from which the mortar falls, form happy hunting-grounds for fern-gatherers. Such a bridge as we have just represented as spanning the beautiful Plym, or one like that we give below, at Dolgelly, is just the kind of structure

for several kinds of ferns to grow on. On the top and sides would be found the Common Polypody, small on the open face of the structure, larger in places where ivy-roots keep in the moisture and retain the leaf-mould. Hartstongues, too, only the smaller specimens, but larger or smaller according to just the same conditions as those which influence the Polypodies, would be found in similar positions. Tiny specimens might be found, too, of the Hard and Soft Prickly Shield Ferns. But

old stony structures are almost the favourite habitats of
the Common and some of the rarer Spleenworts. The
Rue-leaved, the Common Maidenhair, the Black Maidenhair, and the Scaly Spleenworts, are certain to be found
in such places, some in one and some in another, and,
not impossibly though rarely, the Rock, the Alternated,
the Forked, and, if near the sea, the Lanceolate and the
Sea Spleenworts.

Talking of *Asplenium lanceolatum*, let us illustrate one
of its favourite habitats by just this little view of rock.

Our readers will notice the almost perpendicular crevices
in this rock. If within the influence of the sea, this is
just the kind of rocky fissure in which to peer carefully
for the Lanceolate Spleenwort, especially if from above
a tiny rill flows along the rocky surface, and down
between the crevices. In these leaf-mould will gather,
and the air will always be moist, and hence the love for
it of our moisture-loving fern. If such rocks looked
right upon the sea, and were near the beach, then in the
same crevices one might expect to find *Asplenium*

marinum. But the mere mention of that very beautiful glossy-fronded member of the flowerless family brings

sweet Devon again to the mind's eye ; and for the reader who, not having seen it, cannot recall its lovely scenes

to mind, we will give this little peep (page 32) of a cove
in Torbay; and we do this, not only because the peep
itself will be refreshing, but because we can thus illustrate
the habitats of two beautiful ferns. In the lower crannies
of the cliffs, if moisture chances either to be trickling
down from above over the rocky face or oozing out from

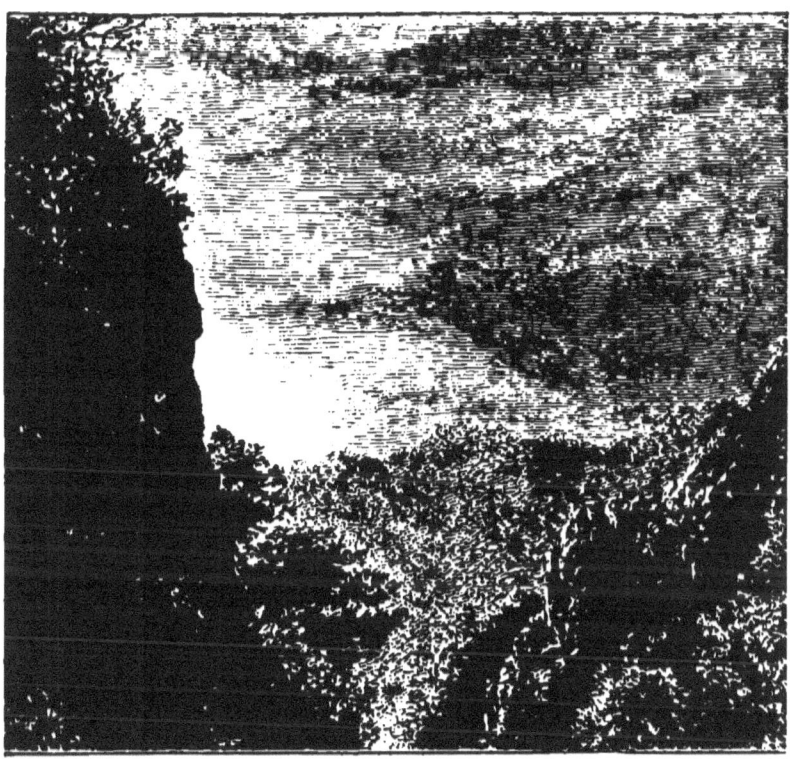

the rock itself, you will be very likely to find *Asplenium
marinum*, and in amongst the shrubs on the overgrown
face of the cliff on the near side is just the kind of place
to hunt for the rare and delicately-beautiful True Maidenhair (*Adiantum capillus-veneris*). These particular cliffs
might not furnish either of the ferns we have mentioned;
but, nevertheless, the places illustrated are just the places

to search in, and the Author has found both ferns in
Torbay.

And now, reluctantly leaving Devonshire and its ferny
scenes, let us illustrate some fern habitats in other places.
And, first, a view shall be given of the far-famed Cheddar
Cliffs (page 33), an especial haunt of the Limestone
Polypody, which grows, as the True Maidenhair is also
asserted to grow, in the moist picturesque nooks of
this rocky region. Rich as it is in many other of the
common kinds of fern, the Cheddar district of Somersetshire must be especially remembered for the two species
just mentioned.

In the succeeding pages the reader will often be told
of rocky habitats for such of the rarer ferns as the
Woodsias, the Holly Fern, the Bladder Ferns, the
Spleenworts, the Rigid Buckler Fern, and the rarer
Polypodies. Here (page 35) is such a one, and, should
the fern-hunter be in any part of the country where, as
the succeeding lists will tell him, he may hope to find
some of these rarer ferns, let him not neglect to carefully
search such likely spots. It would be really difficult
for any one with a real eye for ferns, to pass without
peering into all moist crannies of such rocks, where
" something green " suggests a ferny presence, without a
most careful scrutiny.

On page 36 is yet another bit of suggestive rock. To
climb it may be difficult; yet a jutting fragment here and
there, for the feet to safely secure a hold, and a friendly
shrub growing out from the cliff-side will often tempt one
to climb, if only a little way, to get at some very graceful-
looking clump, that certainly must be a fern of some
kind, and that may chance (who knows?) to be a rare
find, unseen or unexamined by all previous passers-by.

So much for the dry rocky places beloved of the
ferns. Now for the moister ones. There is a species
of dry eloquence in rocks everywhere. But they seem
to *speak* when the mountain torrent rushes over them.

Yet, like nature everywhere, even in this seemingly fierce aspect, there is an under-tone of pathos and tenderness; for how otherwise could the tender and beautiful ferns cling so lovingly to their rough sides?

Let us look at the bit of scenery on the opposite page from "the bonny Dee." In such a neighbourhood

as this we should look and be disappointed not to find the Parsley Fern, all the Polypodies, the Hard Prickly Shield Fern, and the Holly Fern, the Brittle and the Mountain Bladder Ferns, the Male, the Broad and the Mountain Buckler Ferns, the Alternated, the Rue-leaved, the Black Maidenhair, the Green, and the Common Maidenhair Spleenworts.

The mention of Scottish scenery reminds us of a charming picture, in a charming book,—"Habbies

Howe," (page 43), in Dr. Green's "Scottish Pictures." By the courtesy of the publishers, the Religious Tract

Society, we reproduce from their engraving a little bit
of water and fern.   What a happy figure is this fern, and
who can doubt that it is the beautiful Lady Fern? What
beauty and grace does this lovely plant fling over the
wild yet romantic scene in which it figures !

For a bit, now, of characteristic Cumberland scenery,
how could we do better than give the " Lodore Fall ? "
(page 39).   A glance at our Index of Localities at the

end of this volume will tell the reader that Lodore Fall
is one of the habitats of the One-sided Filmy Fern ; but
further search through the succeeding pages will prove
that, out of the forty-five species of ferns figured and
described, no less than thirty-four are to be found in
Cumberland.   Rich, indeed, are the counties of Devon
and Cumberland in the beautiful denizens of wood, lane,
and stream-bank ; and no hunting collector would go
away from such a neighbourhood as that just illustrated

without getting a rich store in numbers and variety of
the flowerless plants.

Whilst we are talking of numbers of this delightful

family that more than others seek the immediate neighbourhood of water, yet without dipping into it, let us
not forget one that alone of its British congeners grows,

not near, but *in* water. We refer to the Marsh Buckler
Fern ; and here below is a tiny bit of "locality" just
suited to *Lastrea thelypteris*. It is obviously boggy, and
in entering at the point shown in the foreground of
the little sketch one would necessarily have to pick
one's way. A bog overgrown with trees, just as this
seems to be, is the place to find the finest specimens of
this water-loving or liquid-peat-soil-loving fern.

Mr. Boot knows how to draw trees and ferns, as may
be seen by the little picture on page 41 of oaks at

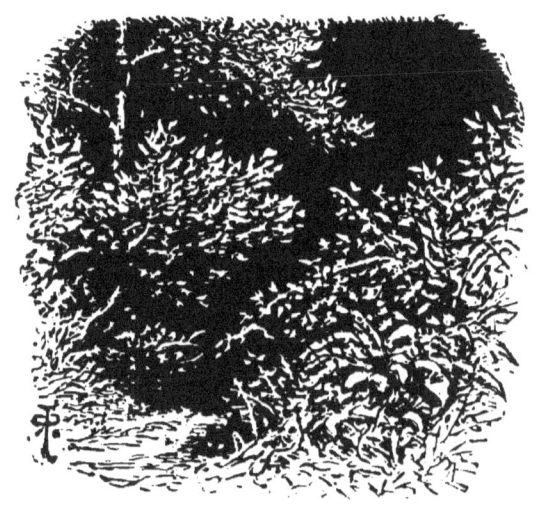

Bradgate ; but the artist leaves the fern-lover to guess
whether he is depicting Bracken or Buckler Fern in the
foreground of his drawing. In just such positions one
might expect to find either the Bracken or the Common
and Broad Buckler Ferns, whilst by the water's edge
there might surely be some Lady Ferns.

Our artists in general have sadly neglected the ferns,
and, when it is considered how much beauty is lent to
all scenery by the presence of ferns, the omission is
strange. On page 42, however, is a happy little sketch by

one who, when he has determined to give us a bit of water,—this is a Severn scene,—and banks sloping down to it, does not forget the important suggestiveness of a few Bracken in the foreground. The depicters of our English scenery can scarcely afford to overlook the Bracken, because it is such a conspicuous feature in all

forest scenes. Mr. Walter Crane understands this, and he knows, too—because he knows the New Forest so well—what Bracken can be in that rich domain. A sketch of his, representing yews and whitebeams in Sloden (page 38), will not be without its especial interest for all who revere and love the magnificent woodlands of

Hampshire. Those who may be tempted to wander that way may like to know that they will, at least, find (besides *Pteris aquilina*), in the

New Forest, *Osmunda regalis, Lastrea montana, Lastrea dilatata, Lastrea filix-mas,* and *Lastrea spinulosa, Polystichum aculeatum,* and *Polystichum angulare, Asplenium adiantum - nigrum,* and the Rue-leaved, the Scaly, and the Common Maidenhair Spleenworts, the Lady Fern, the Hartstongue, the Common Polypody, and the Common Adderstongue.

And now, dear fernhunter, if there be one thing more than another that will stir your enthusiasm, perhaps it will be the sight of a wood such as the one on page 44; for, perhaps, if one place be more fascinating for the fernlover than another, it is the shady, the mysterious, the always delicious depths of a wood when the summer sunshine glints through the trees, bringing up into fine relief the

contorted arms of ancient oaks, on whose ample forks
the Polypody flaunts its golden fruit, and under whose

friendly shade, in the darker and moister angles of the
woodland, Bracken and Buckler Fern display their
graceful forms.

IV.—ON THE CULTIVATION OF FERNS.

HIS chapter is intended to be short and simple in character. Yet it is hoped that its counsel will be none the less useful and effectual.

It is somewhat rare, the Author believes, to find, amongst the numerous valuable and useful works that deal with the home cultivation of plants, books that endeavour to make their instructions relate to the natural conditions under which the same plants were found growing previously to what may be called their domestication. Yet most of our methods of cultivation are but adaptations of natural circumstances, and,—at least in the case of ferns newly gathered from their native habitats,—the closer such natural circumstances or conditions of growth are followed, the more certain will be the success of the adapter ; for it is ignorance of the natural habits of ferns that leads to the most deplorable failures of the growers in pots, rockeries, or cases, of these beautiful, graceful, and interesting plants.

Hence a careful study of the paragraphs which are headed—under the name of each fern, described in these pages,—"Habitats," will throw much more light on the subject of cultivation than the most elaborate but merely routine directions for mixing particular soils.

The natural food of all ferns is leaf-mould, or humus, which is the aggregation in the form of earth of decayed vegetable matter. This is a fact which must be carefully

E 2

borne in mind in considering the economy of fern life. But this fern food must be supplied under certain essential conditions. There must be moisture and more or less of shade, and, with regard to the great majority of ferns, this moisture must be supplied in a particular way. The soil which contains the leaf-mould must be more or less porous, or at least of such a consistency that no stagnant moisture can rest about the roots of the ferns.

If these general circumstances are borne in mind, there will be no difficulty in understanding what follows, or in adapting them to the particular needs of particular ferns ; and, to give a general and comprehensive review of the subject, we will take the ferns in the order in which they are mentioned in the succeeding chapters of this volume, dealing with them singly or in groups.

The Bracken is a free-growing, deeply-rooting fern, flourishing in the open sunshine where the soil is deep and rich, but loving most the shade of woods, because, under trees, the soil is both richer and moister than upon forest glades. Hence a shady or half-shady position upon rich, deep, moist soil will suit this handsome fern.

In its wild state the Hartstongue is so bold and hardy that it will grow almost anywhere, but it especially loves stony habitats, and is small or large according as the stones or rocks from between which it sends up its fronds overlie shallow or deep masses of leafy soil, and are exposed to sunshine and a free circulation of air, or are immersed in shade in a moist atmosphere. Shaded rockery over deep soil is the best position, therefore, for Hartstongues.

The Lady Fern, the Hard Fern, and the Royal Fern, though sometimes found in sunny positions, revel most in soil that is soft, spongy, and rich, and in such positions as secure to them shade and moist air. By fountains or running water will suit them best where their fronds can come within the influence of the spray, and the points of their roots touch the stream without being immersed in it.

So moist is both the actual position,—adjacent to oozing or trickling water,—and the atmosphere surrounding the True Maidenhair and the Annual Maidenhair, that nothing short of the protection of glass will suffice for their successful cultivation ; and for the former the soil should be an extemporisation of the limestone rock and leaf-mould and rocky detritus, out of and in which the Maidenhair naturally grows, whilst for the latter the imitation of the soft, rich soil of its native shady and dripping hedge-bank will suit it best.

Soft leaf-soil under shady rocks best pleases the Wild Parsley Fern, and a rockery habitat of as nearly a similar kind as possible in the garden will meet its home requirements. The only substitute for the dark and dripping caverns, and the moist and shaded rocky crevices where the Bristle Fern grows, is a close covering of glass that excludes the outward air, and rich, sandy, leafy soil; and just such conditions as these are what the Filmy Ferns require, for their natural haunts are similar to those of *Trichomanes radicans.*

Moonwort and Adders-tongue seem to need the companionship, for some mysterious reason, of grassy roots, and, therefore, they should be taken up from their native homes with the grass surrounding them, and the attention of the cultivator must be directed as much to the grassy accompaniments as to the ferns themselves, that they may be kept fresh and healthy.

All the Polypodies love best moist leaf-soil, amongst rocks; and the garden rockery, or the rockery of the fern-case, is the place for them.

The Shield Ferns confess the ferny love for leaf-mould, but they like to toy with the sunshine, and hence they are, perhaps, of all ferns placed in the garden, the most hardy and bold, for they will thrive almost anywhere, and survive adverse conditions that would kill many or their congeners.

Shady rocks with leaf-soil, too, the Bladder Ferns need in their wild homes, and just such conditions will suit them under culture.

The same may be said of the Woodsias, and then we come to the Buckler Ferns, which differ amongst themselves in habit and character. All of them best like the shade, and a rich, porous leaf-mould soil, but only great shade and moisture will suit the Crested and Prickly-toothed Ferns, whilst the Marsh Buckler Fern must grow in as well as on the water.

All the Spleenworts are rock-loving ferns; but the Lanceolate and the Sea Spleenworts cannot grow, out of doors, away from the sea's influence, and, hence, away from the sea, must be put under glass as the only substitute for their natural condition. The Green Spleenwort needs similar treatment, to extemporise the state of saturation of the atmosphere, which it must have for preservation in health and vigour. But the rest of the Spleenworts will grow out of doors on sheltered rockery, if planted firmly and carefully in the crevices between the stones.

Briefly stated, these are the requirements of ferns grown at home.

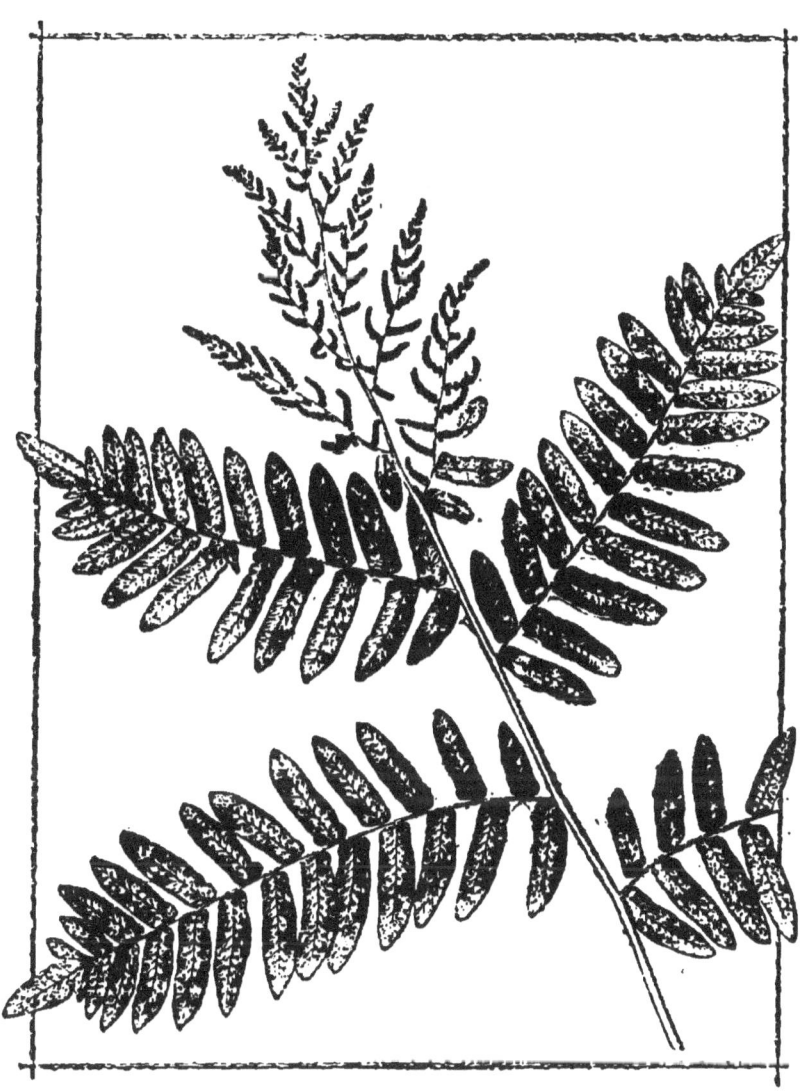

PLATE I.

ROYAL FERN (*Osmunda regalis*). (Fertile frond,)

V.—THE BRACKEN.

Pteris aquilina.

(Plate XI., Fig. 1, page 69.)

LENGTH OF FROND.—One foot to twelve feet, according to the more or less favourable conditions of
growth. The maximum and minimum lengths given
are both exceptional; for, as ordinarily seen, this fern is
from two to six feet long.

GENERAL DESCRIPTION.—*Roots* few in number,
fibrous, but somewhat fleshy, attached, along its entire
length, to the rhizoma. *Rootstock*, a rhizoma—brownish-
black in colour, soft, and thickly covered with short
hair—extending itself both horizontally and perpen-
dicularly; sometimes penetrating to a depth of more
than a dozen feet. *Fronds* deciduous, ordinarily triangular in shape, the leafy part about twice the length
of the stipes: bipinnate in small specimens; tripinnate
in larger ones. The tripinnate may be said to be the
normal form. Pinnæ, placed in nearly opposite pairs
along the rachis, and more or less acutely lance-shaped;
pinnules acutely lance-shaped, pinnate in the lower part
(of tripinnate fronds), pinnatifid higher up, and more or
less entire at the frond apex. Lobes oblong and blunt-
pointed. Towards the apex of the frond the pinnules
are dwindled to mere lobes; nearer it the pinnæ are
also lobe-like, and a lobe terminates the frond. Lobes
concave on their undersides. *Fructification* marginal,
the lines of spore-cases being enclosed in double
indusia formed by elongations or distensions of the
cuticle or membranous surface of the lobes.

HABITATS.—Open commons, downs, and heaths;
glades; woods; hillsides and streamsides; hedgebanks

PLATE II.

BROAD BUCKLER FERN (*Lastrea dilatata*

and fields ; islets in midstream. The Author has occa-
sionally found small specimens growing on the damp
sides of walls, but such a position is only possible for
seedling or very diminutive specimens. The Bracken
frequently covers large spaces of ground, which it ex-
clusively occupies.

WHERE FOUND.—The great abundance of the Bracken
renders it unnecessary to give a detailed list of the
localities in which it grows. The published records of
its distribution, given in the second and revised edition (1883) of Mr. Watson's "Topographical Botany,"
include every county in England, Wales, and Scotland,
except. Wigtonshire and West Ross ; but it is possibly
to be found also in these districts. It grows at various
heights, extending to two thousand feet above the sea-
level.

---◦◦◦---

VI.—THE HARTSTONGUE.

Scolopendrium vulgare.

(Plate VI., Fig. 1, page 59.)

LENGTH OF FROND.—Extremely variable : a couple of
inches when growing on hard, dry walls, to three feet
when in very moist and congenial positions. Ordinary
lengths within these extremes.

GENERAL DESCRIPTION.—*Roots* numerous, fibrous
and somewhat wiry. *Rootstock*, a tufted cormus, the
crown of which is raised slightly above the ground.
Fronds numerous, evergreen, produced in tufts, tongue-
shaped, entire, leathery and glossy, each stipes—about
one-third the length of the leafy part—usually covered
by rust-coloured scales, which often extend along the

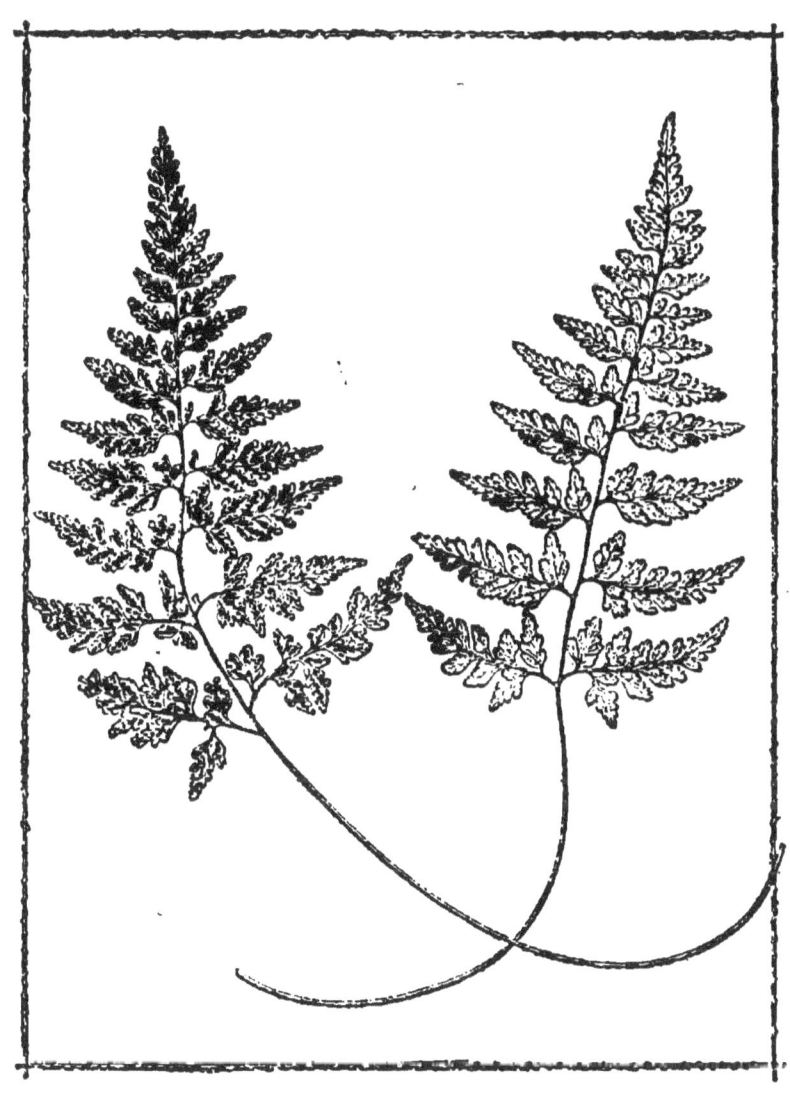

PLATE III.

BLACK MAIDENHAIR SPLEENWORT (*Asplenium adiantum-nigrum*). (Upper and Under Sides.)

under sides of the rachis. Apex of leafy part more or less pointed; base, heart-shaped with ear-shaped projections. *Fructification* produced in parallel lines, which run obliquely from near the rachis towards the leafy margins on either side of the rachis. Each apparent line of spore-cases consists in reality of twin, elongated sori placed side by side and confluent, the scaly indusium, which covers the whole, splitting along the centre when the spores are ripe, and disclosing the densely-clustered, rich-brown spore-cases underneath.

HABITATS.—Shady parts of woods; the bases, sides, and tops of hedgebanks. This species is oftentimes very luxuriant under the shelter of the vegetation of the hedgetop, where it grows frequently in semi-darkness. It grows upon banks overhanging streams; upon rocks and stonework, including walls of buildings and enclosures, bridge-arches, ruins, and the sides of old wells; also upon cliffs overhanging the sea, always, when on stony habitats or elsewhere, most luxuriant where water is oozing or trickling over the rocks, or ground, on which it grows.

WHERE FOUND.—In *England*, in all the counties. In *Wales*, in the counties of Anglesea, Brecknock, Caermarthen, Caernarvon, Denbigh, Flint, Glamorgan, and Pembroke. In the Isle of Man. In *Scotland*, in the following counties:—Aberdeen, Argyle, Ayr, Berwick, Caithness, Dumfries, Inverness, Edinburgh, Elgin, Fife, Forfar, Kincardine, Kirkcudbright, Lanark, Orkney (including the Shetland Isles), Perth, Renfrew, Roxburgh, Selkirk, Stirling, and Sutherland. Also in Cantyre and the Clyde Isles. In *Ireland*, in the Isle of Wight, and in the Channel Islands throughout,—the moisture of the climates of those countries inducing a luxuriant growth of this species, which is found at all elevations up to six hundred feet above the sea-level.

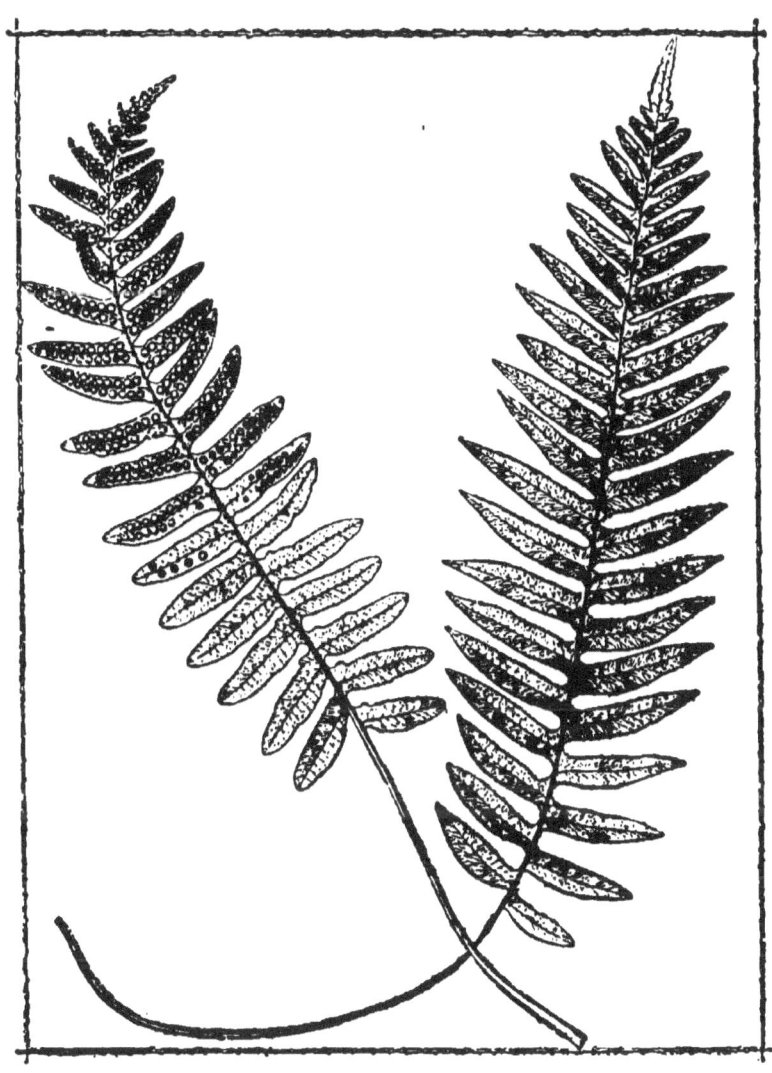

PLATE IV.

COMMON POLYPODY (*Polypodium vulgare*).
(Upper and Under Sides.)

VII.—THE LADY FERN.

Athyrium filix-fœmina.

(Plate VIII., Fig. 1, page 63.)

LENGTH OF FROND.—A foot to five feet, according to position and conditions of growth—largest in the most moist and shady places.

GENERAL DESCRIPTION. — *Roots* fibrous, abundant. *Rootstock*, a tufted cormus, its crown raised slightly above the surface of the ground. *Fronds* numerous, deciduous, delicate, brittle, drooping, produced in tufts. Each stipes usually much shorter than the leafy part, and light green or purplish in colour, with a few scales scattered upon it near the base ; leafy part lance-shaped somewhat broadly ; bipinnate, the pinnæ narrowly lance-shaped and tapering, and placed along the rachis alternately or in opposite pairs ; pinnules blunt-pointed, oblong, serrated, or indented—most deeply near the frond base, less deeply higher up. *Fructification* produced in double rows of sori, one on either side of the midvein of each pinnule, each row of sori being about equidistant from the midvein and the edge of the pinnule. The sori are covered by kidney-shaped indusia, which burst and fall away on the ripening of the spores, whose cases are then light brown in colour.

HABITATS.—The dampest and shadiest parts of woods, especially luxuriant where water oozes over gently-sloping ground ; hedgebanks, in shady lanes ; moist and shady crannies of rocks ; the shady margins of streams, and the sides of waterfalls.

WHERE FOUND.—In *England*, in the counties of Bedford, Berks, Buckingham, Cambridge, Chester, Cornwall, Cumberland, Derby, Devon, Dorset, Durham,

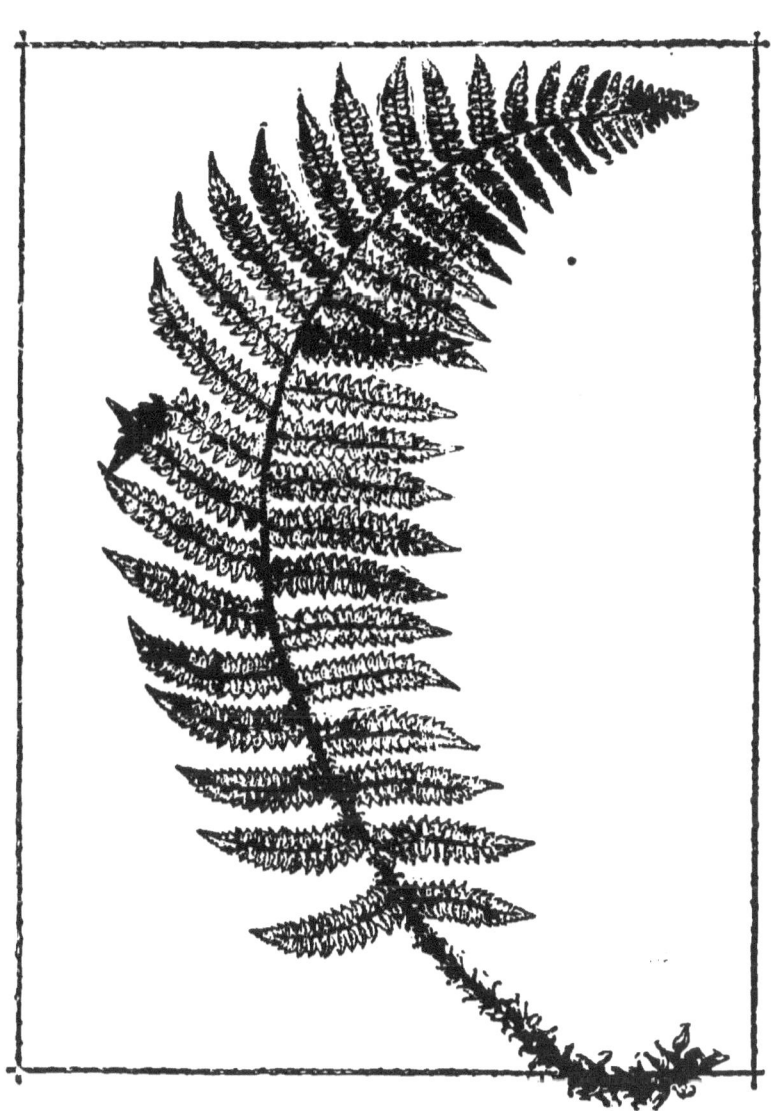

PLATE V.

SOFT PRICKLY SHIELD FERN (*Polystichum angulare*).

Essex, Gloucester, Hants (the mainland and the Isle of Wight), Hereford, Hertford, Kent, Lancaster, Leicester, Lincoln, Middlesex, Monmouth, Norfolk, Northampton, Northumberland, Nottingham, Oxford, Rutland, Salop, Somerset, Stafford, Suffolk, Surrey, Sussex, Warwick, Westmoreland, Wilts, Worcester, and York. In *Wales*, in the counties of Anglesea, Brecknock, Caermarthen, Caernarvon, Cardigan, Denbigh, Flint, Glamorgan, Merioneth, Montgomery, and Pembroke. Specimens have also been found in Radnor. In the Isle of Man. In *Scotland*, in the counties of Aberdeen, Argyle, Ayr, Banff, Berwick, Bute, Caithness, Clackmannan, Cromarty, Dumbarton, Dumfries, Edinburgh, Elgin, Fife, Forfar, Haddington, Inverness, Kincardine, Kinross, Kirkcudbright, Lanark, Linlithgow, Nairn, Orkney, Peebles, Perth, Renfrew, Roxburgh, Selkirk, Stirling, and Sutherland; also in the isles of Arran, Cantire, Harris, Islay, Lewis, and North Uist. In *Ireland*, in the counties of Antrim, Clare, Cork, Dublin, Galway, and Kerry; also in King's County, Kilkenny, Killarney, Limerick, Louth, Waterford, and Wicklow. In the Channel Islands, Jersey and Guernsey. It has been found growing up to two thousand two hundred feet above the sea-level.

—◆◆◆—

VIII.—THE HARD FERN.

Blechnum spicant.

(Plate VI., Figs. 4 and 5, page 59.)

LENGTH OF FROND.—Barren fronds, six inches to two feet; fertile fronds, a foot to three feet—according to the circumstances of growth.

GENERAL DESCRIPTION.—*Roots* wiry, fibrous, abundant. *Rootstock* somewhat thick, creeping, and in time

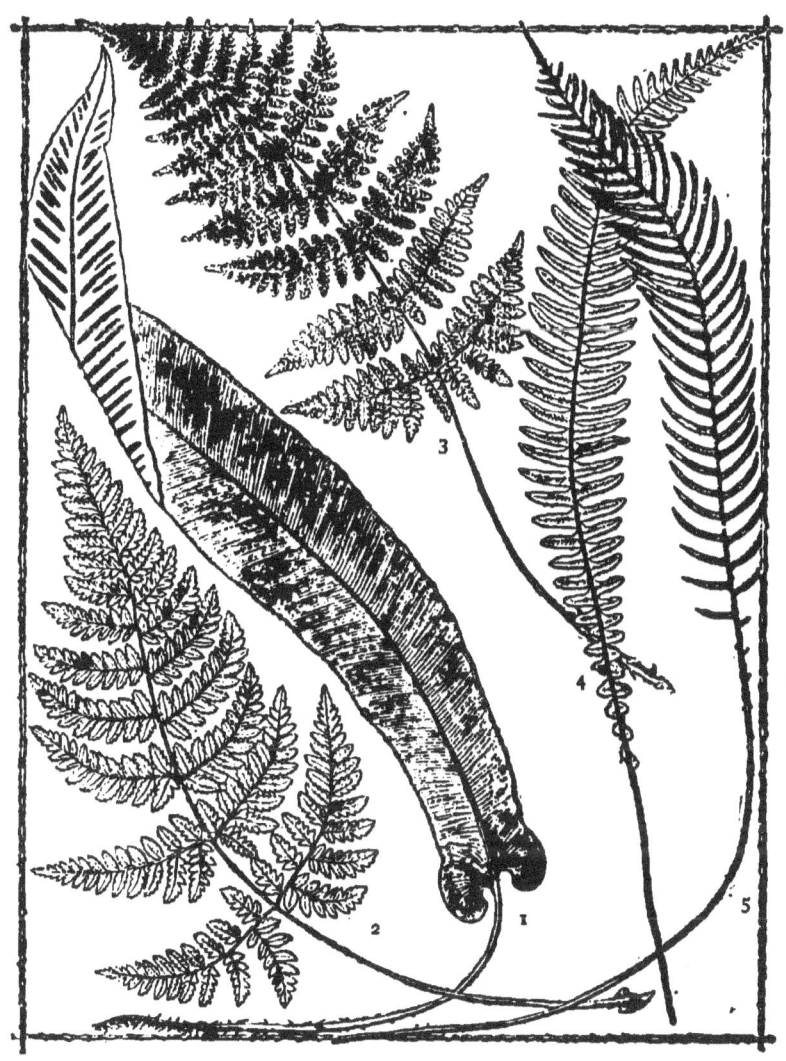

PLATE VI.

1. HARTSTONGUE (*Scolopendrium vulgare*) (Upper and Under Side). 2. HAY-SCENTED BUCKLER FERN (*Lastrea recurva*) (Upper Side). 3. RIGID BUCKLER FERN (*Lastrea rigida*) (Under Side). 4. HARD FERN (*Blechnum spicant*) (Barren Frond, Upper Side). 5. HARD FERN (*Blechnum spicant*) (Fertile Frond, Under Side).

F

forming a cluster of crowns that are consequently attached to each other, the crowns being raised slightly above the ground. *Fronds* numerous, leathery, upper-sides glossy, produced in tufts, and of two kinds—barren and fertile. Barren fronds evergreen, narrowly lance-shaped, tapering at both ends, pinnatifid—sometimes pinnate in their lower parts ; pinnæ narrowly oblong, blunt-pointed, attached by the whole widths of their bases to the rachis, produced in opposite pairs or alter-nately along on each side of the rachis ; stipes reddish-brown, smooth, wiry, from one-fourth to one-seventh the length of the leafy part. Fertile fronds much taller than barren ones, deciduous ; stipes one-third and sometimes one-half the length of leafy part ; leafy part lance-shaped, distinctly pinnate ; pinnæ long, narrow, attenuated, drawn out to a point, in opposite pairs or alternately placed along the rachis and curved upwards. *Fructifi-cation* on fertile fronds only ; sporangia arranged in double lines, one on each side of midvein of each pinna, at first distinct from each other, afterwards becoming confluent, and densely covering the under sides of the pinnæ. The sporangia are covered by elongated indusia, which burst, when the spores are ripe, on the sides next the midveins, and, when thrown back, the spore-cases present a dense, rich-brown mass, ordinarily hiding the whole of the under sides of the pinnæ.

HABITATS.—Moist slopes of woods; damp, stony crevices on hillsides and moorland heights ; stream-margins ; the sides and bases of hedgebanks, especially hedgebanks partly constructed of loose stones ; the stony bases of roadside hedges ; the drier parts of bogs and marshland ; the bases of clumps of shrubbery in forest and woodland glades, and moist nooks of all kinds of rocks, especially in the lowest, most moist, and shady positions.

WHERE FOUND.—In *England*, in the counties of Bedford, Berks, Bucks, Cambridge, Chester, Cornwall,

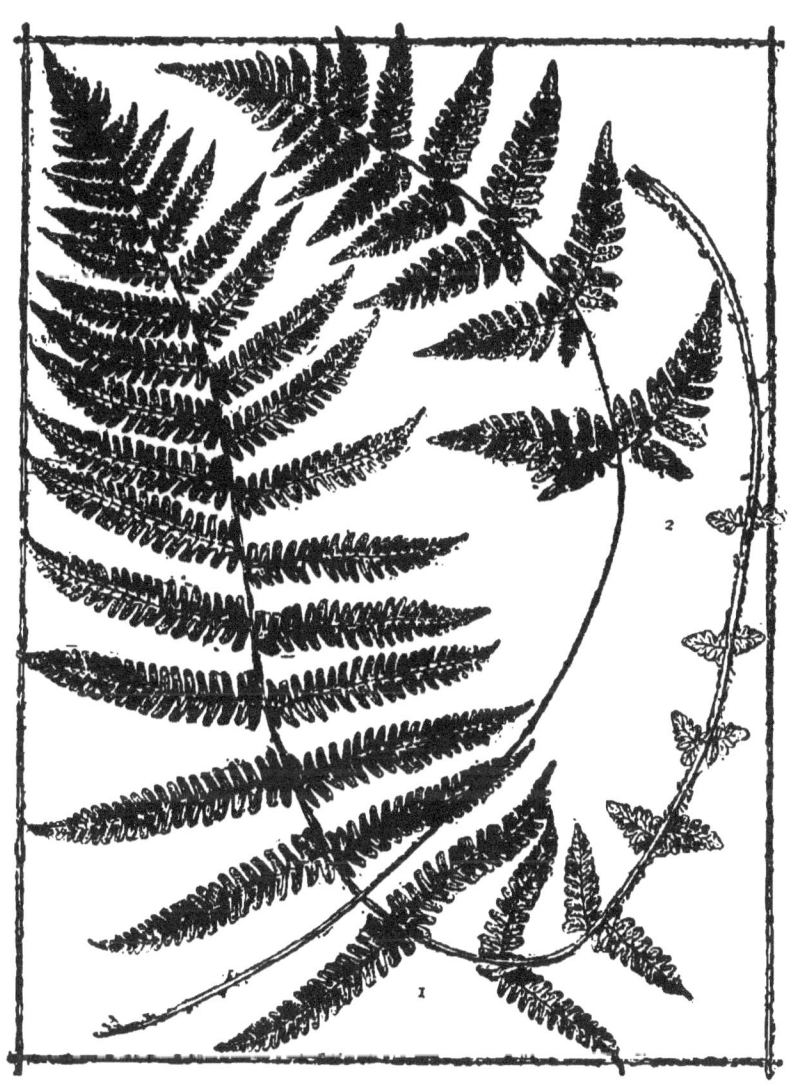

PLATE VII.

1. MOUNTAIN BUCKLER FERN (*Lastrea montana*) (Under Side). 2. PRICKLY-TOOTHED BUCKLER FERN (*Lastrea spinulosa*) (Under Side).

F 2

Cumberland, Derby, Devon, Dorset, Durham, Essex, Gloucester, Hants (the mainland and the Isle of Wight), Hereford, Hertford, Kent, Lancashire, Leicester, Lincoln, Middlesex, Monmouth, Norfolk, Northampton, Northumberland, Nottingham, Oxford, Rutland, Salop, Somerset, Stafford, Suffolk, Surrey, Sussex, Warwick, Westmoreland, Wilts, Worcester, and York.   In *Wales*, in the counties of Anglesea, Brecknock, Caermarthen, Caernarvon, Cardigan, Denbigh, Flint, Glamorgan, Merioneth, Pembroke, and Radnor.   In the Isle of Man.   In *Scotland*, in the counties of Aberdeen, Argyle, Ayr, Banff, Berwick, Bute, Caithness, Clackmannan, Cromarty, Dumbarton, Dumfries, Edinburgh, Elgin, Fife, Forfar, Haddington, Inverness, Kincardine, Kinross, Kirkcudbright, Lanark, Linlithgow, Nairn, Orkney (including the Shetland Isles), Peebles, Perth, Renfrew, Ross, Roxburgh, Selkirk, Stirling, and Sutherland; also in the Isles of Arran, Cantyre, Harris, Islay, Lewis, and North Uist.   In *Ireland*, in the counties of Antrim, Clare, Cork, Down, Dublin, and Galway (the mainland and the Arran Isles); also in King's County, Limerick, Mayo, Tipperary, Waterford, and Wicklow.   In the islands of Jersey and Guernsey.   It ascends to a height of two thousand feet above the sea-level.

IX.—THE ROYAL FERN.

Osmunda regalis.

(Plate I., page 49.)

LENGTH OF FROND.—Two feet to twelve feet, according to more or less congenial conditions of growth; moist peat soil and a boggy situation in immediate contiguity to water favouring and inducing the larger growths.

GENERAL DESCRIPTION.—*Roots* numerous, fibrous,

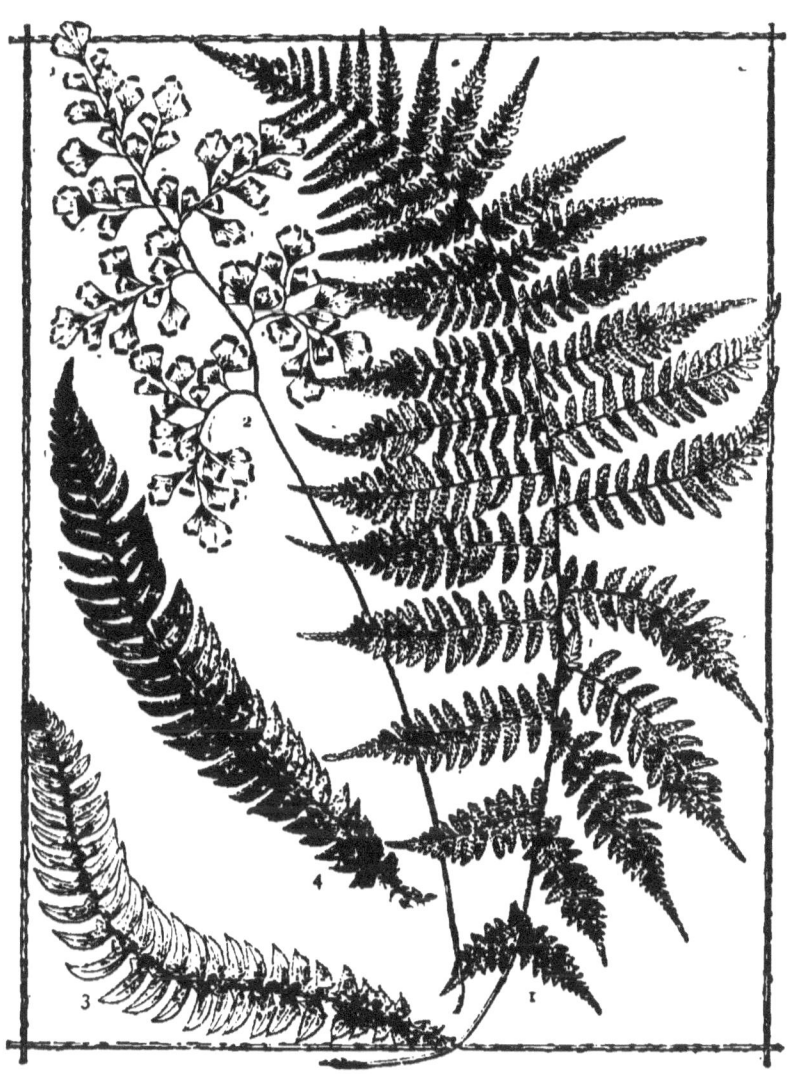

PLATE VIII.

1. LADY FERN (*Athyrium filix-fœmina*) (Under Side). 2. TRUE
MAIDENHAIR (*Adiantum capillus-veneris*) (Under Side).
3. HOLLY FERN (*Polystichum lonchitis*) (Upper Side).
4. HOLLY FERN (*Polystichum lonchitis*) (Under Side).

long, and wiry. *Rootstock*, a thick, tufted cormus; large, in proportion to the size of the plant, and prolonged into a visible, prominent, and above-ground *stem*, raised sometimes to a height of two feet in large-sized, mature plants. The rootstock of a fern, even when not conspicuously raised above the soil, is really its stem, although it does not, in such cases, convey the popular idea of one. The stem of *Osmunda regalis* really becomes, when of large size, a *trunk*, and thus more nearly than any other British species approaches the form and character of a tree-fern. *Fronds* of two kinds —barren and fertile—not very numerous, deciduous, robust-looking, golden green, broadly lance-shaped; very distinctly bipinnate, pinnæ lance-shaped, usually placed in opposite pairs, though sometimes alternately, upon the rachis; pinnules an inch, more or less, in length, oblong, blunt-pointed, in opposite pairs or alternately placed upon the secondary rachides or midstems of the pinnæ. In the fertile fronds the upper pinnæ of the fronds have their pinnules contracted to bear the spores. Stipes about as long as the leafy part. *Fructification* usually, but not always, confined to the upper parts of the fertile fronds, where the pinnules are contracted and bear the globular spore-cases densely crowded upon their under sides—so much so frequently, that the pinnules appear like spikes of inflorescence of a rich, yellowish-brown colour.

HABITATS.—Banks of rivers and lakes, especially in positions close enough to the stream-edge to allow of the roots touching the water; marshy and boggy places, especially where the soil consists largely of peat; low-lying islets, which are sometimes covered densely by little else than plants of this species; damp, low-lying parts of woods; the low-lying parts of moorlands upon ground made marshy by the oozing of water from the heights above; damp meadows and fens, or other peaty places periodically submerged.

PLATE IX.

1. European Bristle Fern (*Trichomanes radicans*) (Upper Side).   2. Limestone Polypody (*Polypodium calcareum*) (Under Side).   3. Three-branched Polypody (*Polypodium dryopteris*) (Under Side).   4. Mountain Polypody (*Polypodium phegopteris*) (Under Side).

WHERE FOUND.—In *England*, in the counties of Bedford, Berks, Bucks, Cambridge, Chester, Cornwall, Cumberland, Devon, Dorset (the mainland and the Isle of Purbeck), Durham, Essex, Hants (the mainland and the Isle of Wight), Hereford, Kent, Lancaster, Leicester, Lincoln, Middlesex, Monmouth, Norfolk, Northumberland, Nottingham, Oxford, Rutland, Salop, Somerset, Stafford, Suffolk, Surrey, Sussex, Warwick, Westmoreland, Wilts, Worcester, and York. In *Wales*, in the counties of Anglesea, Brecknock, Caermarthen, Caernarvon, Denbigh, Flint, Glamorgan, Merioneth, and Pembroke. In the Isle of Man. In *Scotland*, in the counties of Aberdeen, Argyle, Ayr, Berwick, Caithness, Clackmannan, Dumbarton, Dumfries, Fife, Forfar, Haddington, Kincardine, Kirkcudbright, Lanark, Linlithgow, Orkney (including the Shetland Isles), Perth, Renfrew, Ross, Stirling, Sutherland, and Wigton. In the isles of Arran, Bute, Harris, Islay, Lewis, Mull, and North Uist. In *Ireland*, in the counties of Clare, Cork, Donegal, Dublin, Galway, and Kerry; also in King's County, Mayo, Tipperary, Waterford, and Wicklow. In Jersey. *Osmunda regalis* grows at various altitudes up to a thousand feet above the sea-level.

———◦◦◦———

X.—THE TRUE MAIDENHAIR.

Adiantum capillus-veneris.

(Plate VIII., Fig. 2, page 63.)

LENGTH OF FROND.—Six inches to two feet, according to position and other circumstances of growth; but the maximum length given is exceptional.

GENERAL DESCRIPTION.—*Roots* black, fibrous, somewhat fleshy. *Rootstock*, a creeping rhizoma, slender, covered with black scales, and extending itself along the

PLATE X.

1. MARSH BUCKLER FERN (*Lastrea thelypteris*) (Barren Frond, Upper Side).   2. BRITTLE BLADDER FERN (*Cystopteris fragilis*) (Under Side).   3. MOUNTAIN BLADDER FERN (*Cystopteris montana*) (Upper Side).   4. SEA SPLEENWORT (*Asplenium marinum*) (Upper Side).   5. SEA SPLEENWORT (*Asplenium marinum*) (Under Side).

surface of the rock or soil upon which the plant is growing—the roots underneath holding it in position. *Fronds* triangular, numerous, evergreen, delicate, usually tripinnate, but sometimes only bipinnate. In the tripinnate fronds the pinnæ are mostly triangular, but are sometimes variously shaped, and are divided into pinnules, which, near the bases of the pinnæ, are again divided into distinctly-stalked, fan-shaped, more or less cleft or indented, lobes. Towards the apices of such divided pinnæ the pinnules are not again divided, but are simply stalked and indented. In all compound ferns there is always less division, both of fronds, pinnæ, and pinnules, towards the apex of each frond, pinna and pinnule. Stipes, usually about the same length as the leafy part, purplish black, smooth, and shining. Rachis and secondary rachides purplish black, shining, and hairlike. *Fructification* marginal, produced at the outer and upper edges of the under sides of the fertile lobes, and consisting of oblong sori, covered by indusia formed by the reflexed and blanched margins of the lobes.

HABITATS.—Cliffs at or near the sea-coast—seldom inland. The moist hollows and crannies of limestone rocks are the favourite habitats of this species. It should be looked for in sea-caverns; under rocky ledges or spurs ; in semi-dark crevices, and behind or under the shadow of cliffside bushes or scrub. Very often it is completely hidden by a screen of bushes or other vegetation on the face of rocks—in such positions growing almost in darkness. Frequently it grows on inaccessible parts of steep cliffs ; but whenever rocks are searched for specimens, those especial nooks moistened by oozing or trickling streams of water, flowing down or along the rocky surface, should be carefully examined.

WHERE FOUND.—In *England*, in the counties of Cornwall, Devon, Dorset, Salop, and Somerset only ; the particular localities in those counties being the fol-

PLATE XI.

1. BRACKEN (*Pteris aquilina*) (Upper Portion of Frond, Upper Side). 2. MOONWORT (*Botrychium lunaria*). 3. ADDERS-TONGUE (*Ophioglossum vulgatum*). 4. LITTLE ADDERS-TONGUE (*Ophioglossum lusitanicum*).

lowing : in Cornwall, on cliffs at Carclew, at Carrick
Gladden (on the sea-coast between Hayle and St. Ives),
and upon cliffs at Penzance; in Devonshire, near
Brixham (upon the limestone rocks of Mewstone Bay),
on cliffs at Ilfracombe, and also at Watermouth, near
Ilfracombe ; in Shropshire, at Titherston Clee Hill ; in
Somersetshire, on the Cheddar Cliffs and on the coast at
Clevedon.   In *Wales*, in the county of Glamorgan, on
the coast at Dunraven, on Barry Island, at East
Aberthaw, and at Port Kirig.   In the Isle of Man,
between Douglas and Peel, and in Glen Meay.   In
Scotland, in the county of Kincardine, on the banks of
the river Carron.   In *Ireland*, in the counties of Clare,
Galway, and Kerry : in the first-named county at Bally-
vaughan, or between that place and Cremlin Point; in
Kerry, at Cahir Conree near Tralee ; and in Galway, at
Lough Bulard, near Urrisberg, and at Roundstone,
Connemara : also in the Arran Isles.   On cliffs in
Jersey and Guernsey *Adiantum capillus-veneris* has also
been found.

XI.—THE ANNUAL MAIDENHAIR.

Gymnogramma leptophylla.

(Plate XII., Figs. 3 and 4, page 71.)

LENGTH OF FROND.—Three to nine inches.
GENERAL DESCRIPTION.—*Roots* fibrous. *Rootstalk*
small, tufted. *Fronds* annual, deciduous ; stipes from
one-third to one-half the length of leafy part, dark
brown at the base and green above ; the first fronds
shorter than the later ones and simply pinnate, the
pinnæ borne on short stalks alternately on each side of
the rachis—fan-shaped and indented.   The taller and

PLATE XII.

1. HARD PRICKLY SHIELD FERN (*Polystichum aculeatum*) (Upper Side). 2. ALPINE BLADDER FERN (*Cystopteris regia*) (Under Side). 3. ANNUAL MAIDENHAIR (*Gymnogramma leptophylla*) (Upper Side). 4. ANNUAL MAIDENHAIR (*Gymnogramma leptophylla*) (Under Side). 5. PARSLEY FERN (*Allosorus crispus*) (Barren Frond). 6. PARSLEY FERN (*Allosorus crispus*) (Fertile Frond).

later fronds bipinnate, sometimes tripinnate, the pinnæ
ovate and alternate, and bearing fan-shaped, indented,
alternate pinnules.   The shape of the pinnules very
much resembles that of the lobes of the True Maidenhair.   *Fructification* non-indusiate, consisting of sori
arranged in lines at the backs of the pinnules, but often
becoming confluent.

HABITATS.—The most shady and sheltered sides of
hedgebanks.   It grows oftentimes amongst other dwarf
vegetation, especially in places where water trickles or
oozes over the banks.

WHERE FOUND.—No reliable evidence has been produced as to the finding of this little fern in any other
part of the British Islands than Jersey, in some localities
of which—such as St. Aubin, St. Haule, and St.
Laurence—it grows in abundance.   But it is quite
possible, we think, that diligent search might lead to
this pretty little fern being found somewhere along the
south coast of England.

XII.—THE MOUNTAIN PARSLEY FERN.

Allosorus crispus.

(Plate XII., Figs. 5 and 6, page 71.)

LENGTH OF FROND.—Barren fronds four to eight
inches; fertile fronds six to twelve inches, according to
more or less congenial conditions of growth.

GENERAL DESCRIPTION.—*Roots* numerous, fibrous,
wiry, often matted into a dense mass.   *Rootstock* thick,
tufted, often elongating into numerous crowns.   *Fronds* of
two kinds, both produced in dense, tufted clusters.   Barren
fronds, bright green, triangular, bipinnate, and sometimes
tripinnate ; pinnæ triangular, opposite or alternate; pin-

PLATE XIII.

1. ALPINE POLYPODY (*Polypodium alpestre*) (Upper Side). 2. LANCEOLATE SPLEENWORT (*Asplenium lanceolatum*) (Upper Side). 3. LANCEOLATE SPLEENWORT (*Asplenium lanceolatum*) (Under Side). 4. SCALY SPLEENWORT (*Asplenium ceterach*) 5. SCALY SPLEENWORT (*Asplenium ceterach*) (Under Side). 6. ROCK SPLEENWORT (*Asplenium fontanum*) (Upper Side). 7. ROCK SPLEENWORT (*Asplenium fontanum*) (Under Side). 8. RUE-LEAVED SPLEENWORT (*Asplenium ruta-muraria*) (Upper Side). 9. RUE-LEAVED SPLEENWORT (*Asplenium ruta-muraria*) (Under Side).

nules wedge-shaped, alternate on opposite sides of the secondary rachides; lobes—in the tripinnate form—club-shaped or wedge-shaped, and indented upon their margins; stipes about equal in length to the leafy part, green, and brittle. Fertile fronds are similar in general arrangement of parts to barren fronds, but the ultimate divisions are contracted into oblong, rounded, spore-bearing lobes. The stipes of each fertile frond is frequently three times as long as the leafy part. *Fructification* borne upon the whole of the under sides of the lobes of the fertile fronds, the edges of which are rolled under so far as to meet, thus enclosing the spore-cases in simple indusia and giving a rounded form to each lobe. When ripe the lobes and their contents turn brown and open to allow of the escape of the spores.

HABITATS.—Moist crevices of rocks; spaces between loose stones upon hillsides.—*Allosorus crispus* sometimes in such positions growing in great abundance. So thickly are plants of this species often clustered that they have obtained the common name of "Rock Brakes."

WHERE FOUND.—In *England*, in the counties of Chester, Cumberland, Derby, Devon, Durham, Hereford, Lancaster, Northumberland, Salop, Somerset, Westmoreland, Worcester, and York. In *Wales*, in the counties of Anglesea, Caernarvon, Cardigan, Denbigh, Glamorgan, Merioneth, Montgomery, and Radnor. In the Isle of Man. In *Scotland*, in the counties of Aberdeen, Argyle, Ayr, Berwick, Caithness, Dumbarton, Dumfries, Elgin, Fife, Forfar, Inverness, Kincardine, Kinross, Kirkcudbright, Lanark, Peebles, Perth, Renfrew, Ross, Roxburgh, Selkirk, Stirling, and Sutherland. In the isles of Arran, Harris, Mull, and Skye. In *Ireland*, only in the counties of Antrim, Clare, Down, and Louth. In these four counties the localities are believed to be very few in which *Allosorus crispus* has been found. They are as follows:—In Antrim, at Carrickfergus; in Clare, at Blackhead; in Downshire,

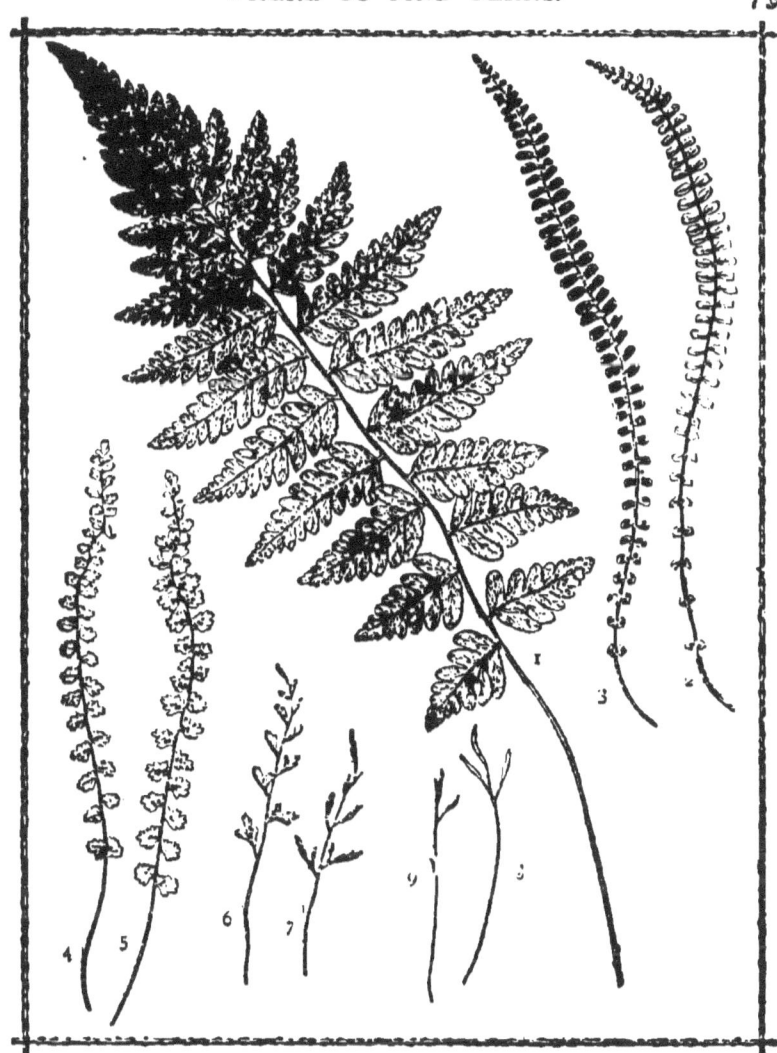

PLATE XIV.

1. CRESTED BUCKLER FERN (*Lastrea cristata*) (Upper Side)
2. COMMON MAIDENHAIR SPLEENWORT (*Asplenium tricho-manes*) (Upper Side). 3. COMMON MAIDENHAIR SPLEEN-WORT (*Asplenium trichomanes*) (Under Side). 4. GREEN SPLEENWORT (*Asplenium viride*) (Upper Side). 5. GREEN SPLEENWORT (*Asplenium viride*) (Under Side). 6. ALTERNATE SPLEENWORT (*Asplenium germanicum*) (Upper Side). 7. ALTERNATE SPLEENWORT (*Asplenium germanicum*) (Under Side). 8. FORKED SPLEENWORT (*Asplenium septentrionale*) (Upper Side). 9. FORKED SPLEENWORT (*Asplenium septentrionale*) (Under Side).

G

at Sleive Bignian and on the Mourne Mountains;
and in Louth, on the Carlingford Mountains. It grows at
heights reaching to three thousand five hundred feet
above the sea-level.

XIII.--THE BRISTLE FERN.

Trichomanes radicans.

(Plate IX., Fig. 1, page 65.)

LENGTH OF FROND.—Six inches to a foot and a half.
GENERAL DESCRIPTION. —*Roots* fibrous, blackish,
woolly, and numerous. *Rootstock*, a creeping rhizoma—
black and covered with scales—that extends itself along
upon the surface of the rocks upon which it is found
growing. *Fronds* evergreen, triangular, tripinnate; stipes
—about equal in length to the leafy part or less—purplish
black, as also are the rachides. Pinnæ triangular and
alternate upon the rachis; pinnules 'ovate or lance-
shaped, alternate upon the secondary rachides—lobes
irregularly-shaped, but somewhat oblong, alternate, and
deeply incised or serrated. Leafy, narrow wings run along
on either side of the stipes, rachis, and secondary rachides.
General character of the leafy texture of the frond pellucid.
Fructification in urn-shaped receptacles produced near
the ends of veins projected—bristle-like—beyond the
lobe-margins, and through and beyond the urn-shaped
receptacles.
HABITATS.—The wet sides of rocks and caves where
the most absolute shade prevails and the air is laden
with reeking moisture. Such habitats are essential to
the very life of this beautiful fern, whose pellucid texture

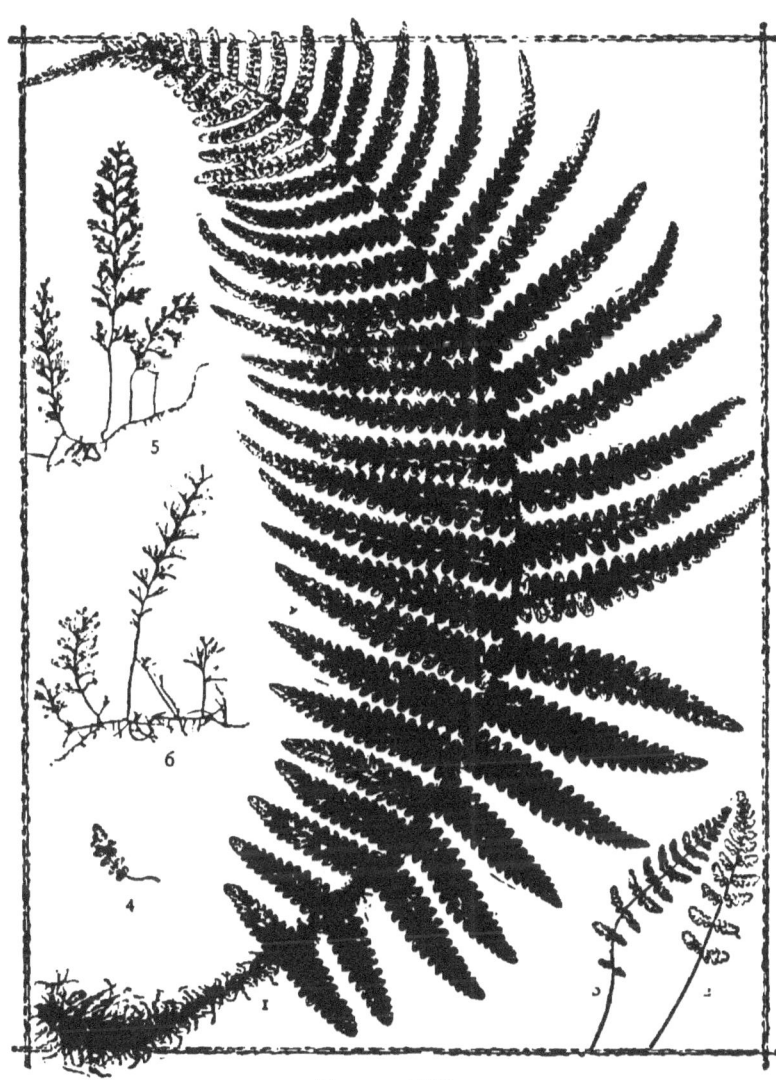

PLATE XV.

1. MALE FERN (*Lastrea filix-mas*) (Under Side).   2. OBLONG
WOODSIA (*Woodsia ilvensis*) (Upper Side).   3. OBLONG
WOODSIA (*Woodsia ilvensis*) (Under Side). 4. ALPINE WOODSIA
(*Woodsia alpina*) (Under Side).   5. TUNBRIDGE FILMY FERN
(*Hymenophyllum tunbridgense*).   6. ONE-SIDED FILMY FERN
(*Hymenophyllum unilaterale*).

would shrivel up under the effects of sunshine or of a
dry atmosphere.

WHERE FOUND.—No locality in either *England* or
Scotland is at present generally known to possess this
fern, although it is said to have been found in Cornwall
and West Yorkshire, in Arran and Argyle. It is believed that it grows abundantly in one part of *North
Wales* (Merioneth is the county which has been named),
but the locality is only known to a few persons, who
have kept its whereabouts a profound secret. In *Ireland*, it has been found in the counties of Cork, Kerry,
Limerick, Waterford, and Wicklow; and in the following
localities: in Cork county, in Glendine wood; near
Youghal, at Glenbour and Killeagh; on rocks near
Bandon; on rocks at Ballinasy Glen and Temple Michael
Glen near Cork; at the Clashgariffe Fall; on rocks near
Glandore, and also near Bantry; and on rocks on Carri-
geena Kildorrery in the north of Cork. In Kerry County,
on the Tork Mountains and at the Tork Waterfall;
amongst the Cromaglaun Mountains; at Glengariff in
Valentia Island; near Dingle (on Mount Eagle); at
Gortgaree, between Killarney and Kenmare; at Black-
stones, in Glouin Caragh; and at Inveragh and Curaan
Lake, Waterville. In County Limerick, amongst the
Cumailte Mountains. In County Waterford, along the
Blackwater Valley; and in Wicklow County, at Powerscourt Waterfall and in Hermitage Glen. In some of
these localities—the mountainous ones—it has been
found growing at a height of fifteen hundred feet above
the sea-level.

———•◦•———

XIV.—THE MOONWORT.

Botrychium lunaria.

(Plate XI., Fig. 2, page 69.)

LENGTH OF FROND.—Two to ten inches.

GENERAL DESCRIPTION.—*Roots* few in number, thick, and fleshy. *Rootstock* fleshy, small, elongated, erect growing, tuberous. *Fronds* of two parts barren and fertile: the one leafy, the other spore-bearing. A common stipes supports both from the base to about midway on the frond where the leafy portion diverges. It consists of a single, somewhat bluntly lance-shaped pinna, with pairs of opposite or alternate, crescent-shaped, fan-shaped, or half-moon-shaped pinnules. The stipes, or, strictly speaking, the rachis, continuing upwards and beyond the leafy pinna, terminates in a single, branched cluster of spore-cases. *Fructification*—the fruitful part of the frond is simply pinnate or bipinnate, the branches alternate and again alternately branched in its lower part, each branch bearing a small cluster of globular spore-cases, which at the season of ripening turn from the incipient green to a golden-brown colour.

HABITATS.—The open face of heaths, damp meadows, and moors, amongst grass on spots somewhat elevated but not extremely damp.

WHERE FOUND.—In *England*, in the counties of Bedford, Bucks, Cambridge, Chester, Cornwall, Cumberland, Derby, Devon, Dorset, Durham, Essex, Gloucester, Hants (the mainland and the Isle of Wight), Hereford, Kent, Lancaster, Leicester, Lincoln, Monmouth, Norfolk, Northampton, Northumberland, Nottingham, Oxford, Rutland, Salop, Somerset, Stafford, Suffolk, Surrey, Sussex, Warwick, Westmoreland, Wilts, Worcester, and

York. In *Wales*, in the counties of Anglesea, Caer-
marthen, Caernarvon, Denbigh, Flint, Glamorgan,
Merioneth, Montgomery, Pembroke, and Radnor. In the
Isle of Man. In *Scotland*, in the counties of Aberdeen,
Argyle, Ayr, Banff, Berwick, Caithness, Clackmannan,
Dumbarton, Dumfries, Edinburgh, Elgin, Fife, Forfar,
Haddington, Inverness, Kincardine, Kinross, Kirkcud-
bright, Lanark, Linlithgow, Nairn, Orkney (including
Shetland), Perth, Renfrew, Ross, Roxburgh, Selkirk,
Stirling, and Wigton : also in the islands of Bute and
Skye. In *Ireland*, in the counties of Antrim, Cork,
Down, Galway, Londonderry, and Wicklow. It is found
at various heights up to two thousand seven hundred
feet above the sea-level.

XV.—THE ADDERS-TONGUE.

Ophioglossum vulgatum.

(Plate XI., Fig. 3, page 69.)

LENGTH OF FROND.—Three to twelve inches, the
maximum length given being exceptional.

GENERAL DESCRIPTION.—*Roots* few in number, brittle,
thick, fleshy. *Rootstock* erect, elongated, fleshy, small in
size. *Fronds* of two parts, barren and fertile, having a
common stipes—the barren part a single, oval-shaped,
entire pinna (equal in size to the circumference of a hen's
egg), the base of which forms the top of the sheath that
constitutes the upper part of the stipes and clasps the
lower part of the stem of the fertile portion of the frond.
Fructification produced in small globular spore-cases
arranged in two lines, on opposite sides of the upper
part of the fruitful branch of the frond, which thus
becomes a terminal spike of fructification.

HABITATS.—Heaths, moors, pastures, amongst grass-roots in loamy soil, and in damp positions.

WHERE FOUND.—In *England*, in the counties of Bedford, Berks, Bucks, Cambridge, Chester, Cornwall, Cumberland, Derby, Devon, Dorset, Durham, Essex, Gloucester, Hants (the mainland and the Isle of Wight), Hereford, Hertford, Huntingdon, Kent, Lancaster, Leicester, Lincoln, Middlesex, Monmouth, Norfolk, Northampton, Northumberland, Nottingham, Oxford, Salop, Somerset, Stafford, Suffolk, Surrey, Sussex, Warwick, Westmoreland, Wilts, Worcester, and York. In *Wales*, in the counties of Anglesea, Caernarvon, Denbigh, Flint, Glamorgan, Pembroke, and Radnor. In *Scotland*, in the counties of Aberdeen, Argyle, Ayr, Berwick, Clackmannan, Edinburgh, Elgin, Fife, Forfar, Haddington, Kincardine, Kinross, Kirkcudbright, Lanark, Linlithgow, Orkney (including Shetland), Perth, Renfrew, Selkirk, and Stirling. In *Ireland*, in the counties of Antrim, Armagh, Cork, Dublin, Galway, and Tipperary. It is found growing at various heights up to a thousand feet above the sea-level.

XVI.—THE LITTLE ADDERS-TONGUE.

Ophioglossum lusitanicum.

(Plate XI., Fig. 4, page 69.)

LENGTH OF FROND.—Two to four inches.
GENERAL DESCRIPTION.—*Roots* few in number, fleshy, brittle. *Rootstock* small, upright in growth, fleshy, tuberous, elongated. *Fronds* of two parts, barren and fruitful, consisting, upon a common stipes, of a single narrow, entire, somewhat grass-like, barren pinna, and of a taller branch forming a stem in continuation of the

stipes, and bearing at the upper end the fruitful spike.
There is a single barren frond (though sometimes there
may be two barren fronds on the same plant) and a single
fruitful branch, as in the case of *Ophioglossum vulgatum*,
to which fern the present species bears a general, though
diminutive, resemblance. *Fructification* — the fruitful
spike, pointed at the end, consists of two rows, one on
each side of the rachis, of rounded spore-cases—each
row consisting, usually, of about five or six of these
cases.

HABITATS.—Damp positions on heaths and on open
pastures, amongst grass-roots.

WHERE FOUND.—The only locality generally known
is the Island of Guernsey, where, in 1854, it was first
discovered in the neighbourhood of Petit Bot Bay.   It
is said to have been found in Cornwall, and it is quite
possible that, owing to its inconspicuousness, it may
abound in many parts of the British Isles without having
been discovered.

XVII.—THE COMMON POLYPODY.

Polypodium vulgare.

(Plate IV., page 55.)

LENGTH OF FROND.—Two or three inches to two feet
and a half, according to position and circumstances of
growth--the maximum length given being, however,
very exceptional, though fronds of that length have been
found by the Author.   The average size of specimens is
given in most descriptions of ferns at from six to
eighteen inches—the specimens commonly encountered
being seldom more than a foot in length.

GENERAL DESCRIPTION.—*Roots* abundant, fibrous.

Rootstock, a hairy or scaly rhizoma, which branches and creeps in various directions upon the surface of the rock, wall, or soil in which the plant is growing, producing fronds from numerous points of its upper side. *Fronds* evergreen, numerous, deeply pinnatifid, of a somewhat elongated egg-shape, pointed at the apex, and divided into long, blunt-pointed, entire pinnæ, an inch or more in length—each resembling somewhat the finger of the hand—with deep wide clefts between each. Stipes of varying length, green, smooth, brittle, sometimes as long as, sometimes shorter than, and often much longer than, the leafy part. *Fructification* usually confined to the upper portion of the undersides of the pinnæ, consisting of two rows of non-indusiate, rounded sori, one on each side of the midvein of each pinna, generally crowded, and sometimes becoming confluent. When ripened, the sporangia turn to a rich orange, or brown, colour.

HABITATS.—The tops and sides of rocks and walls. It is especially luxuriant where moist seams of earth, lying in shaded positions, afford abundant root-room, and it is oftentimes much stunted and diminutive on the drier, exposed, and sunny faces of rocks and walls. Old walls falling into ruin are always found to have accumulated soil between their loose stones. Should trees be growing around, this accretion of soil will be largely composed of leaf-mould, and upon the shadowy sides of such walls all rock or wall-growing ferns will be found in the greatest state of vigour and luxuriance. The Common Polypody grows also in the forks of old trees where leaf-mould has accumulated; upon tree-stumps raised above, or almost level with, the ground; in the sides and upon the tops of hedgebanks, amongst loose stones, or in the stumps, trunks, forks, or hollows of trees growing in hedgebanks. Pollard-trees in hedgebanks afford favourite habitats of this fern. Old bridge-arches, and indeed all old or decaying stonework, are, similarly, favourable positions for *Polypodium vulgare*. Wherever, in short, leaf-mould has accumulated in stony or woody places, it may be looked for, as its creeping, vigorous rhizomas love to occupy the congenial habitats which shade and a leaf-soil provide.

WHERE FOUND.—In every county of *England, Wales, Scotland,* and *Ireland;* in the Isle of Man, and the Channel Islands, growing in many places in extreme abundance. *Polypodium vulgare, Lastrea filix-mas* (the Male Fern), and *Pteris aquilina* (the Bracken) are the most plentiful and widely-distributed of all British ferns.

XVIII.—THE MOUNTAIN POLYPODY.

Polypodium phegopteris.

(Plate IX., Fig. 4, page 65.)

LENGTH OF FROND.—Six inches to a foot and a half or twenty inches.

GENERAL DESCRIPTION. — *Roots* fibrous, somewhat long, and numerous. *Rootstock,* a rhizoma, slender but vigorous, creeping extensively and horizontally along or just underneath the soil. *Fronds* delicate, herbaceous, abundant, springing from numerous points along the upper sides of the rhizomas ; stipes delicate, pale green, slender, brittle, about double the length of the leafy part; leafy part triangular, often pinnate in its lower part, pinnatifid higher up. Pinnæ ordinarily in opposite pairs and pinnatifid, the pinnules nearest the main rachis being sometimes again pinnate in the lowest pair of pinnæ, which ordinarily hang downwards in a peculiar manner distinct from the others. The form of the pinnæ in the lower part of the frond is somewhat lance-shaped, their bases tapering towards the rachis and their apices drawn out to somewhat acute points. *Fructification* distributed equally over every part of the frond and almost marginal, consisting of two rows of non-indusiate, rounded sori, one on each side of the midvein of each pinnule.

HABITATS.—Damp woods in mountainous districts, or in country that is more or less hilly ; the margins of mountain or moorland streams ; the immediate vicinity of waterfalls, in the fine spray of which this beautiful species delights ; moist nooks in rocks, especially in the neighbourhood of water. The habitats of *Polypodium phegopteris* are essentially shady.

WHERE FOUND.—In *England*, in the counties of Chester, Cornwall, Cumberland, Derby, Devon, Dorset, Durham, Gloucester, Hereford, Lancaster, Monmouth, Northumberland, Salop, Somerset, Stafford, Sussex, Westmoreland, Wilts, and York. In the Isle of Man. In *Wales*, in the counties of Brecknock, Caermarthen, Caernarvon, Cardigan, Denbigh, Glamorgan, Merioneth, Montgomery, and Radnor. In *Scotland*, in the counties of Aberdeen, Argyle, Ayr, Banff, Berwick, Caithness, Clackmannan, Dumbarton, Dumfries, Edinburgh, Elgin, Fife, Forfar, Inverness, Kincardine, Kinross, Kirkcud-bright, Lanark, Linlithgow, Orkney (including Shetland), Perth, Renfrew, Ross, Roxburgh, Selkirk, Stirling, and Sutherland ; also in the isles of Cantyre, Islay, and Mull. In *Ireland*, it is found only in the counties of Antrim, Donegal, Down, Galway, Kerry, Londonderry, Louth, and Wicklow. It is found growing at various heights up to three thousand five hundred feet above the sea-level.

XIX.—THE THREE-BRANCHED POLYPODY.

Polypodium dryopteris.

(Plate IX., Fig. 3, page 65.)

LENGTH OF FROND.—Six to twelve inches.
GENERAL DESCRIPTION.—*Roots* fibrous, delicate, numerous. *Rootstock*, a slender, somewhat black rhizoma,
which creeps extensively, in a horizontal direction, upon
or just underneath the soil. *Fronds* triangular, abundant,
springing from numerous points of the rhizomas, delicate,
brittle, golden green, herbaceous, each with a slender,
brittle, pale-green stipes and a three-branched leafy part,
about half the length of the stipes ; rachis and rachides
also very slender, delicate, and brittle.   The branches of
the frond grow at right angles to each other, and each is,
itself, more or less triangular in shape, with a clear space of
stem between it and the point of attachment to the rachis.
The two lower branches are ordinarily pinnate at and near
the base and pinnatifid higher up, and are divided into
pairs of oblong, more or less deeply-indented pinnules,
the lower ones (near the main rachis) of each pair being
longer than the upper cnes.   The upper branch is
divided into opposite pairs of more or less deeply-cleft
pinnæ, which become gradually merged into each other
towards the apex of the branch that forms the apex of
the frond.   *Fructification* produced in rows of non-
indusiate sori, one row on each side of the midvein of
pinnule or pinna, according to the size and development
of the plant.
    HABITATS.—Slightly less moist than those of *Polypo-
dium phegopteris :* shady woods amongst underwood and
in rocky crevices ; streamsides and shady hedgebanks in
hilly, moorland, or mountainous districts.

WHERE FOUND.—In *England*, in the counties of Chester, Cornwall, Cumberland, Derby, Devon, Durham, Gloucester, Hereford, Lancaster, Lincoln, Monmouth, Northumberland, Oxford, Salop, Somerset, Stafford, Sussex, Warwick, Westmoreland, Worcester, and York. In *Wales*, in the counties of Anglesea, Brecknock, Caernarvon, Cardigan, Denbigh, Flint, Glamorgan, Merioneth, Montgomery, Pembroke, and Radnor. In *Scotland*, in the counties of Aberdeen, Argyle, Ayr, Banff, Berwick, Caithness, Clackmannan, Dumbarton, Dumfries, Edinburgh, Elgin, Fife, Forfar, Inverness, Kincardine, Kinross, Kirkcudbright, Lanark, Linlithgow, Nairn, Peebles, Perth, Renfrew, Ross, Roxburgh, Selkirk, Stirling, and Sutherland. In the isles of Arran and Mull, and in Shetland. In *Ireland*, only in the counties of Antrim, Down, Galway, and Kerry. It is found growing up to and at elevations of two thousand seven hundred feet above the sea-level.

XX.—THE LIMESTONE POLYPODY.

Polypodium calcareum.

(Plate IX., Fig. 2, page 65.)

LENGTH OF FROND.—Eight to eighteen inches.
GENERAL DESCRIPTION.—*Roots* black, numerous fibrous. *Rootstock*, a rhizoma branched, slender, black, extensively creeping. *Fronds* abundant, triangular, bluish-green, produced from numerous points of the upper sides of the rhizomas; less fragile than those of *Polypodium dryopteris ;* stipes of varying lengths, generally longer than the leafy part, pale green, bipinnate, and, in large and luxuriant specimens, tripinnate ;

pinnæ in pairs or alternate upon the rachis ; lowest pair of pinnæ somewhat narrowly triangular, pinnate and sometimes bipinnate at their bases, and divided into opposite or alternate, oblong, and somewhat cone-shaped pinnules, which are more or less deeply cleft into narrow, oblong, blunt-pointed lobes. The succeeding and upper pairs of pinnæ are less and less divided, on the same principle, as they near the apex of the frond, the divisions of the several pinnæ being similarly less and less towards their apices. This principle of gradation is always observed in all ferns—the divisions or indentations of all the parts of fronds being less and less from base to apex of frond, pinna, pinnule, lobe, or serrature. As in *Polypodium dryopteris,* the lower pairs of pinnæ have their lower pinnules longer and more developed than those on the upper sides of their respective secondary rachides. *Fructification* marginal on the lobes or pinnules—according to the size and development of the pinnæ—and bearing a strong general resemblance to the arrangement of the fructification of the Bracken. But in *Polypodium calcareum* the sporangia are non-indusiate. The fructification is spread equally over the whole under surface of the frond, the pinnules of which are concave on their under sides, giving to them a curled, crisped appearance.

HABITATS.—Limestone districts, in moist crevices of limestone rocks and amongst loose limestones. It prefers shady positions especially for its roots, but it will oftentimes be found growing in places that are somewhat sunny.

WHERE FOUND.—In *England,* in the counties of Bucks, Cumberland, Derby, Durham, Gloucester, Hereford, Lancaster, Oxford, Salop, Somerset, Stafford, Westmoreland, Wilts, Worcester, and York. In *Wales,* in the counties of Brecknock, Caernarvon, Denbigh, and Glamorgan. In *Scotland,* it is said to have been found growing wild in the counties of Aberdeen and Perth, but it is believed to be extremely rare in that country. In *Ireland*, no plants of this species have been found. In Britain, *Polypodium calcareum* grows at various heights up to twelve hundred feet above the sea-level.

———◦◇◦———

XXI.—THE ALPINE POLYPODY.

Polypodium alpestre.

(Plate XIII., Fig. 1, page 73.)

LENGTH OF FROND.—One foot to three feet and a half.

GENERAL DESCRIPTION.—*Roots* fibrous, abundant. *Rootstock*, a cormus, erect, short, tufted. *Fronds* broad, lance-shaped, drawn out to a point at the apex, and considerably narrowed towards the base ; bipinnate ; pinnæ alternate on the rachis, narrow, pointed, symmetrical, divided into pairs of oblong, somewhat blunt-pointed, and deeply-indented pinnules. Stipes short, with a few light-coloured scales scattered upon it. This fern was for a long time confounded with the Lady Fern, *Athyrium filix-fœmina*, but it was distinguished from that species and included amongst the Polypodies in 1841, having been discovered in that year by Mr. Hewett C. Watson. *Fructification* distinct from *Athyrium filix-fœmina*, and consisting of round, non-indusiate sori usually produced in two rows along each pinnule, a sorus being placed ordinarily in those parts of the pinnules immediately contiguous to the bases of the notches between the lobes.

HABITATS.—Shady rocks and streamsides ; often covering considerable spaces of ground in mountainous districts.

WHERE FOUND.—Only in *Scotland*, in the counties of Aberdeen, Argyle, Banff, Forfar, Inverness, Perth, and Sutherland, occurring at elevations reaching from twelve hundred to three thousand six hundred feet above the sea-level, in company with, and in similar positions to, *Athyrium filix-fœmina* until the highest range of that species is reached, when *Polypodium alpestre* occurs alone in the higher elevations.

<hr>

XXII.—THE HARD PRICKLY SHIELD FERN.

Polystichum aculeatum.

(Plate XII., Fig. 1, page 71.)

LENGTH OF FROND.—One to four feet.

GENERAL DESCRIPTION.—*Roots* long, fibrous, tough, abundant. *Rootstock*, a large, tufted cormus, the crown of which is raised above the ground. *Fronds* lance-shaped, leathery in texture, dark green, produced in a circle around the crown, which, with the short stipes, is thickly covered with rust-coloured or reddish-brown scales that are usually thickly scattered upon the rachis and also upon the secondary rachides. Leafy part of frond bipinnate ; pinnæ alternate, lance-shaped, divided into alternate, wing-shaped, serrated, and bristly pinnules, attached by their bases, more or less narrowed, to the secondary rachides or midstems of the pinnæ. The pinnules, separate and distinct from each other at the inner ends of the pinnæ, are decurrent or merged into each other at their bases, towards and at the apices of the pinnæ. The upper pinnule on each pinna situated next the principal rachis is larger than any of the others on the same pinna, and its apex sometimes overlaps the

base of the pinnule next above it. *Fructification* produced in rows—cne on each side of the midvein of
each pinnule, or, towards the apex of the frond and
towards the apex of the pinna, on each side of the
midveins of the pinnæ themselves—of round sori,
covered by round indusia.

HABITATS.—The sloping ground of woods where
shaded by trees or dwarfer growths; the sides of hedge
and other embankments which make the boundaries of
shady lanes; the sides of hills, especially where fragments of rock and sheltering shrubs cover ground
enriched by leaf-mould. Dwarf specimens or seedlings
may sometimes be found upon walls; but such positions
are exceptional, as only depths of rich earth can afford
the root-room required by large and luxuriant plants of
this species.

WHERE FOUND.—In *England*, in the counties of
Bedford, Berks, Bucks, Cambridge, Chester, Cornwall,
Cumberland, Derby, Devon, Dorset, Durham, Essex,
Gloucester, Hants (the mainland and the Isle of Wight),
Hereford, Hertford, Kent, Lancaster, Leicester, Lincoln,
Middlesex, Monmouth, Norfolk, Northampton, Northum-
berland, Nottingham, Oxford, Salop, Somerset, Stafford,
Suffolk, Surrey, Sussex, Warwick, Westmoreland, Wilts,
Worcester, and York. In *Wales*, in the counties of
Anglesea, Brecknock, Caermarthen, Caernarvon, Den-
bigh, Flint, Glamorgan, and Pembroke. In the Isle of
Man. In *Scotland*, in the counties of Aberdeen, Argyle,
Ayr, Berwick, Caithness, Clackmannan, Dumfries,
Edinburgh, Elgin, Fife, Forfar, Haddington, Inverness,
Kincardine, Kinross, Kirkcudbright, Lanark, Nairn,
Orkney, Peebles, Perth, Renfrew, Ross, Roxburgh,
Selkirk, Stirling, and Sutherland : also in the isles of
Bute, Cantyre, and Islay. In *Ireland*, in the counties
of Antrim, Clare, Dublin, Galway, and Wicklow. It
is found in Jersey. It ascends to two thousand five
hundred feet above the sea-level.

XXIII.—THE SOFT PRICKLY SHIELD FERN.

Polystichum angulare.

(Plate V., page 57.)

LENGTH OF FROND.—One to four feet.

GENERAL DESCRIPTION.—*Roots* long, fibrous, abundant. *Rootstock*, a thick, tufted cormus, the crown being raised above the ground. *Fronds* lance-shaped, somewhat soft in texture, light green, sometimes golden green, though at times much darker in colour, produced in a circle around the crown, which with the short stipides— each stipes being about one-fourth the length of the leafy part of the frond—is densely covered with rust-coloured scales. These are continued thickly upon the rachis and also frequently upon the secondary rachides. Leafy part of frond bipinnate, pinnæ alternate, lance-shaped, divided into angular, slightly-indented, and somewhat hairy pinnules, each of which is distinctly stalked, though the stalk is short. The pinnules are alternate upon the secondary rachides. The entire aspect of the fronds of *Polystichum angulare* is more lax and drooping than that of *Polystichum aculeatum*, and the pinnules are more distinctly angular than those of its congener, though in some other respects the two species very much resemble each other. *Fructification* produced in rows of sori, one row on each side of the midvein of each pinnule. The sori are round, and are covered in their early stage by round indusia, which fall off when the ripening of the spores is completed.

HABITATS.—Woods, in all kinds of positions upon the ground, growing oftentimes luxuriantly under trees, or wherever there are rich deposits of leaf-soil ; stream-sides, in the shade ; lanes, upon the sides and tops of hedgebanks ; hillsides, amongst shrubs and broken rocks ; the long, sloping sides of cuttings which border roadsides in hilly country ; and the hedgebanks which run on either side of roadways. *Polystichum angulare* is oftentimes found in great abundance.

WHERE FOUND.—In *England*, in the counties of Berks, Bucks, Chester, Cornwall, Cumberland, Derby, Devon, Dorset, Durham, Essex, Gloucester, Hants (the mainland and the Isle of Wight), Hereford, Hertford, Huntingdon, Kent, Lancaster, Leicester, Middlesex, Norfolk, Northumberland, Salop, Somerset, Stafford, Suffolk, Surrey, Sussex, Warwick, Westmoreland, Wilts, Worcester, and York. In *Wales*, in the counties of Anglesea, Brecknock, Caermarthen, Caernarvon, Cardigan, Denbigh, Flint, Glamorgan, Pembroke, and Radnor. In the Isle of Man. In *Scotland*, only in the counties of Ayr, Argyle, Berwick, and Roxburgh. In *Ireland*, in the counties of Antrim, Clare, Cork, Dublin, Galway, Kilkenny, Tipperary, Waterford, and Wicklow. Also in the Arran Isles. It grows also in Jersey and Guernsey. It is found growing at various heights up to two thousand five hundred feet above the sea-level.

——•◦•——

XXIV.—THE HOLLY FERN.

Polystichum lonchitis.

(Plate VIII., Figs. 3 and 4, page 63.)

LENGTH OF FROND.—Six inches to two feet.

GENERAL DESCRIPTION.—*Roots* fibrous, wiry, tough. *Rootstock*, a tufted, somewhat thick cormus. *Fronds* narrowly lance-shaped, evergreen, rigid, leathery, spiny, simply pinnate, each frond strongly resembling a pinna of *Polystichum angulare.* The serrated, bristly pinnæ are alternate along and on opposite sides of the rachis, and wing-shaped, and are attached to the rachis by their narrowed bases, the upper portion of each pinna next the rachis ordinarily overlapping the base of the pinna next above it; stipes very short and scaly. *Fructification* usually present only on the upper sides of the fronds, and consisting of rows—one on each side of the midvein of each pinna—of round sori, covered, when the spore-cases are young, by round indusia. The sori are usually so arranged that they form an acute angle on the under-side of each fruitful pinna, the angle being at the apex of each pinna, the lines which form it widening out towards the base.

HABITATS.—Mostly in localities not less than a thousand feet above the sea-level; in such localities it grows in moist, rocky fissures, and is oftentimes firmly and immovably wedged into stony crevices.

WHERE FOUND.—In *England*, in the counties of Cumberland, Durham, Northumberland, Westmoreland, and York. The particular localities in three of these counties are the following : in Cumberland, at Fairfield, Helvellyn ; in Durham, on the Falcon Clints, Teesdale, some ten miles westward of Middleton, and also on the Mazebeck Scar ; in the county of York, on Attermire Scar ; in the neighbourhoods of Giggleswick and Ingleborough, and (near Settle) at Langcliffe. In *Wales*, in Caernarvon, Glamorgan, and Merioneth. In Caernarvon, the neighbourhoods of Clogwyn-y-garnedd, of Cwm-Idwal, of Glyder-Vawr, and of Twll-du. In Merioneth, it has been found (on Cader Idris) by Mr. Franklin T. Richards. In *Scotland*, in the counties of Aberdeen, Argyle, Banff, Caithness, Dumbarton, Elgin, Forfar, Inverness, Orkney, Perth, Ross, Stirling, and Sutherland ; also in the Isle of Mull, on Ben More. The special localities of some of these countries are these: in the county of Forfar, on the Clova Mountains, Canlochen, on Craig Maid, in Glen Isla, in Glen Dole, and in Glen Fiadh ; in the county of Inverness, amongst the mountains and rocks near Loch Erricht ; in the county of Perth, on Ben Chonzie, near Crieff, on Ben Lawers, on Ben Voirlich, on Craig Challiach, and in Glen Lyon. In the county of Ross, near Castle Leod, on the Raven Rock ; and in the county of Sutherland, at Assynt and on Ben Hope. In *Ireland*, in the counties of Donegal, Kerry, Leitrim, Meath, and Sligo ; and in the following places : to the east of Lough Eske, in a glen on the Rosses, and in the Thanet mountain passes. In Kerry, on Brandon Hill ; in Leitrim, on the Glenade Mountains ; in Meath, at Navan, and in Sligo, on the Ben Bulben Mountains. The Holly Fern is found at heights ranging from a thousand feet above the sea-level to three thousand two hundred feet above it.

XXV.—THE BRITTLE BLADDER FERN.

Cystopteris fragilis.

(Plate X., Fig. 2, page 67.)

LENGTH OF FROND.—Six to fourteen inches, depending on the character of its habitats.

GENERAL DESCRIPTION.—*Roots* black, fibrous, wiry, numerous. *Rootstock*, a small, tufted cormus, which spreads laterally, forming several adjacent crowns. *Fronds* in numerous tufts from each crown, delicate-green, brittle, herbaceous ; stipes of varying lengths, very brittle ; leafy part broadly lance-shaped, bipinnate, the ovate pinnæ alternate or in pairs along the rachis, and divided into irregularly-alternate, ovate pinnules, which are again divided into rounded, oblong, much-indented lobes. *Fructification* irregularly but abundantly distributed over the under sides of the lobes, and consisting of roundish sori, covered by inflated, bladder-like or hood-like indusia, attached each by one side—that towards the base of the lobe—and falling off when the spores are ripened. The sori then frequently become confluent, and cover the entire under sides of the fronds with their rich-brown fructification.

HABITATS.—Shady and moist crevices of rocks, especially limestone rocks; though, owing to its hardiness, this species may be found in other rocky habitats. Its rootstocks are often so firmly ensconced in the stony chinks it loves best as to render their extraction difficult or impossible; but in other cases, when growing amongst loose stones, it is easily obtainable. It grows also on walls and on stony banks, always preferring their shady sides.

WHERE FOUND.—In *England*, in the counties of Chester, Cornwall, Cumberland, Derby, Devon, Dorset, Durham, Essex, Gloucester, Hereford, Kent, Lancaster, Leicester, Middlesex, Monmouth, Norfolk, Northampton, Northumberland, Nottingham, Salop, Somerset, Stafford, Suffolk, Surrey, Sussex, Warwick, Westmoreland, Wilts, Worcester, and York. In *Wales*, in the counties of Anglesea, Brecknock, Caermarthen, Caernarvon, Cardigan, Denbigh, Flint, Glamorgan, Merioneth, Montgomery, and Radnor. In *Scotland*, in the counties of Aberdeen, Argyle, Ayr, Banff, Berwick, Caithness, Clackmannan, Dumbarton, Dumfries, Edinburgh, Elgin, Fife, Forfar, Haddington, Inverness, Kincardine, Kinross, Kirkcudbright, Lanark, Linlithgow, Nairn, Orkney, Perth, Renfrew, Ross, Roxburgh, Selkirk, Stirling, and Sutherland: also in the Hebrides. In *Ireland*, in the counties of Antrim, Cork, Down, Galway, Kerry, Leitrim, Sligo, and Wicklow. It is found growing at various heights up to nearly four thousand feet above the sea-level.

XXVI.—THE ALPINE BLADDER FERN.

Cystopteris regia.

(Plate XII., Fig. 2, page 71.)

LENGTH OF FROND.—Four to ten inches.
GENERAL DESCRIPTION. *Roots* fibrous, black, wiry,
numerous. *Rootstock*, a small, tufted cormus. *Fronds*
numerous, brittle, herbaceous, delicate, produced in
tufts ; stipes ordinarily short ; leafy portion somewhat
broadly lance-shaped, bipinnate ; pinnæ—in opposite
pairs upon the rachis or alternate—short, ovate, and
again divided into bluntly-ovate, deeply-incised pinnules.
This fern resembles a rounded form of *Cystopteris
fragilis. Fructification* produced abundantly over all
the under-surface of the frond, and consisting of round
sori covered by the hood-like indusia, each sorus keeping itself distinct from the others. Hence the sori of
this species do not become confluent, as frequently do
those of *Cystopteris fragilis.*
HABITATS.—The moist fissures of rocks and the
earthy seams of old walls.
WHERE FOUND.—This fern has been discovered in
very few localities in Britain, though it is quite possible
that it is much more plentiful than is generally supposed.
The places where it has been found growing in *England*
are in the counties of Cumberland, Derby, Durham,
Essex, and York ; at Saddleback in Cumberland, and
at Low Leyton in Essex, in which last-named place it
was found upon an old wall from which it has now dis-
appeared. In *Wales*, it is said to have been found at
Cwm-Idwal and on Snowdon ; and on Ben Lawers in
Scotland.

XXVII.—THE MOUNTAIN BLADDER FERN.

Cystopteris montana.

(Plate X., Fig. 3, page 67.)

LENGTH OF FROND.—Four to ten inches.

GENERAL DESCRIPTION.—*Roots* fibrous, not very abundant. *Rootstock*, a rhizoma, which creeps considerably in a horizontal direction, thin and dark-coloured. *Fronds* abundant, bright green, brittle, herbaceous, produced from numerous points along the rhizoma ; stipes about twice the length of the leafy part, which is somewhat triangular in general form and tripinnate in its lower part, though bipinnate higher up. Pinnæ alternate or opposite, generally the former, on the rachis. The basal pinnules of the two lowest pinnæ are much longer on the lower than on the upper sides of their midstems or secondary rachides, and these elongated pinnules are again divided into alternate, egg-shaped, and deeply-indented lobes, thus becoming tripinnate. The remaining pinnules are less and less divided both towards the apex of the frond and towards the apices of their respective pinnæ. *Fructification* abundant upon the fronds, and consisting of round sori, covered, when young, by the bladder-like or hood-like indusia which are characteristic of the genus *Cystopteris*.

HABITATS.—Rocky fissures in mountainous districts and the rocky margins of mountain streams. Where rich leaf-mould has collected in such fissures, this species grows luxuriantly, always preferring the most complete shade.

WHERE FOUND.—In *Scotland*, only in the counties of Aberdeen, Forfar, and Perth ; the particular districts in the two last-named counties being in Canlochen, at the head of Glen Isla, in Forfarshire, and on Ben Lawers, and at Corrach Dh' Oufillach, between Glen Lochy and Glen Dochart, in the county of Perth. But it is possibly much more abundant than these rare "finds" would seem to indicate.

———•◦•———

XXVIII.—THE OBLONG WOODSIA.

Woodsia ilvensis.

(Plate XV., Figs 2 and 3, page 77.)

LENGTH OF FROND.—Two to six inches.

GENERAL DESCRIPTION.—*Roots* fibrous, wiry. *Rootstock* small, tufted. *Fronds* numerous, brittle, deciduous, thick and woolly in texture, produced in clusters from the crown; stipes of varying lengths, generally rather short, jointed, reddish, breaking off when the fronds begin to decay a little distance above the crown; leafy parts hairy or woolly, oblong, lance-shaped, pinnate; pinnæ opposite or alternate, oblong, egg-shaped, short, pinnatifid, and divided into small, blunt-pointed pinnules, the incisions between which reach down almost to the midstems of the pinnæ. *Fructification* consisting of spore-cases somewhat marginal upon the undersides of the pinnules, and provided with indusia which lie as a sort of scales under the sori, with a fringed margin, which is spread over them. The thickly hair-covered undersurfaces of the pinnules afford a sort of shelter for the sporangia.

HABITATS.—Moist crevices of rocks in mountainous districts at such altitudes as lie between twelve hundred and three thousand feet above the sea-level.

WHERE FOUND.—In *England*, in the counties of Cumberland, Durham, Westmoreland, and York; in Durham, on basaltic rocks in the neighbourhood of Cauldron Snout and on Falcon Clints, Teesdale. In *Wales*, in the counties of Caernarvon and Merioneth ; in Caernarvonshire, at the pass of Llanberis amongst limestone rocks ; also on rocks at Clogwyn-y-Garnedd and in similar positions at the little Dog's Lake (Llyn-y-Cwm) near Glyder Vawr. Here the plants have been reputed to be abundant, but difficult of access, owing to the steepness of the rocks. In *Scotland*, in the counties of Dumfries, Elgin, Forfar, and Perth ; and in the following localities in those counties : in Dumfries, at the " Devil's Beeftub," upon rocks in a ravine near Loch Skene, at a farm called Corehead near Moffatt, and upon hills near Moffatt; also amongst crumbling rocks upon hills dividing Dumfries from Peebles ; in Elgin, near Forres ; in Forfar, in Glen Fiadh amongst the Clova Mountains. It is also found on rocks upon Ben Chonzie, near Crieff, and its other habitats in Perthshire are on Ben Lawers. It has not been recorded as having been found in *Ireland :* but it is quite possibly present in many localities, where it has not been discovered by botanists who are in the habit of publishing their " finds."

———•◊•———

XXIX.—THE ALPINE WOODSIA.

Woodsia alpina.

(Plate XV., Fig. 4, page 77.)

LENGTH OF FROND.—One to six inches.

GENERAL DESCRIPTION.—*Roots* slender, fibrous, wiry. *Rootstock* slender, tufted. *Fronds* small, thick, leathery, hairy—but less hairy than *Woodsia ilvensis*—numerous, produced in tufts from the crown, pinnate, lance-shaped; stipes rather short, slightly hairy; pinnæ short, in pairs .

or alternate, sometimes distant from each other, egg-shaped, and divided into two or three rounded, blunt pinnules, or, in small plants, lobes, the clefts between them being more or less deep according to their size. *Fructification* produced upon the margins of the pinnules, and protected by indusia in the form of scales, which lie under the sori and have fringed margins, which are spread over them, as already indicated in the case of the species last described. But in *Woodsia alpina* the undersides of the pinnules are less hairy than are those of the Oblong Woodsia, and the sporangia are consequently better seen.

HABITATS.—Similar to those of its congener *Woodsia ilvensis*, namely, moist crevices of rocks at altitudes between twelve hundred and three thousand feet above the sea-level.

WHERE FOUND.—Only in *Wales* and *Scotland*. In *Wales*, in the county of Caernarvon only; on the eastern side of Snowdon in a rocky chasm called Clogwyn-y-Garnedd, and on limestone rocks at Moel Lechog at the pass of Llanberis. In *Scotland*, in the counties of Dumfries, Forfar, and Perth—the habitats in Forfarshire being in Glen Fiadh, on the Clova Mountains, and in Glen Isla; and in Perthshire, on Ben Chonzie (near Crieff), on Ben Lawers, at Catiaghiamman, on Craig Challiach, and at Mael-dun-crosk. But, as with other reputedly "rare" ferns, it is quite possible that it is much more plentiful than is generally supposed.

—◦≺◦—

XXX.—The Male Fern.

Lastrea filix-mas.

(Plate XV., Fig. 1, page 77.)

LENGTH OF FROND.—One foot to five feet, according to its more or less congenial conditions of growth.

GENERAL DESCRIPTION.—*Roots* abundant, long, wiry, fibrous. *Rootstock*, a large, tufted cormus, whose crown is sometimes raised several inches above the ground, and is always raised to some extent. *Fronds* broadly lance-shaped, numerous, rigid, thick, bold-looking, somewhat leathery, produced in a circle around the crown, shuttle-cock-shape; stipes usually very short—not exceeding a sixth of the length of the leafy part—densely covered, as is the crown of the rootstock and the under (and sometimes the upper) sides of the rachides, by rust-coloured scales, which often extend in smaller form and less thickly to the under sides of the rachides or mid-stems of the pinnæ; leafy part pinnate in small specimens and partially bipinnate in more luxuriant ones; pinnæ placed on the rachis in opposite pairs, or alternately, long, tapering, and pointed, widest at their bases, becoming smaller gradually outwards, and again divided into oblong, somewhat short, blunt pinnules closely set together with great regularity—so much so that their apices form almost straight lines. These symmetrical pinnæ are either pinnate or pinnatifid—some being the one and some the other in finely-developed specimens—the tendency to division being always less towards the apex of the frond and towards the apices of the pinnæ. *Fructification* usually confined to upper half of under side of frond, and consisting of rows of sori, a row on each side of the midvein of each pinnule—each sorus being covered by a kidney-shaped indusium attached by its notched side, but falling off when the spores are ripe.

HABITATS. — Woods, glades, commons, heaths, streamsides, hillsides, rocks, walls, cliffs, banks and mounds, and green lanes—growing in almost every imaginable position. The ground under trees in woods; sloping ground of open parts of woods or forests ; rocky embankments; the ground under forest undergrowth ; the sides of waterfalls; hedgetops ; hedgesides; ditches where there is motion in the water. This species sometimes grows in the shade, often in the full sunshine—a pigmy when found on walls or other "stony places" where there is no depth of earth—a giant (amongst its kind) when in shadow in a vapour-laden atmosphere and in congenial soil. It grows, in short, almost everywhere.

WHERE FOUND.—In *England, Wales, Scotland, Ireland*, and all the British Isles, large or small, this abundant fern is found. No soil on which fern-life is at all possible is likely to be foreign to *Lastrea filix-mas*. From the sea-level at various altitudes up to two thousand five hundred feet above it, the Male Fern is abundantly distributed.

———

XXXI.—THE BROAD BUCKLER FERN.

Lastrea dilatata.

(Plate II., page 51.)

LENGTH OF FROND.—One to six feet.

GENERAL DESCRIPTION.—*Roots* abundant, fibrous, wiry. *Rootstock*, a large, tufted cormus, its crown raised a little above the surface of the soil. *Fronds* deciduous, produced around the crown, dark green, arching, numerous, broadly lance-shaped, sometimes nearly triangular, tripinnate at the base, and bipinnate above; stipes of varying lengths half as long, a third as long, or the same length as the leafy part, scattered over with dark-coloured scales; pinnæ opposite or alternate along the rachis, narrowly triangular in shape, and divided into oblong pinnules alternate on the secondary rachides, the pinnules being again divided into larger or smaller sharply-incised lobes, whose under sides are concave. The two pinnæ at the base of the frond have the pinnules on the under sides of their midstems longer than those above them, and more developed (being consequently tripinnate). The next pair or two above partake slightly of the same character, and the pinnules gradually become equal on both sides towards the apex of the frond. *Fructification* in rows of small sori, one on each side of each pinnule or lobe, according to the size and development of the pinnæ, scattered pretty evenly over the under surface of the frond, and covered, in its early stage, by kidney-shaped indusia, which fall away when the spores have ripened.

HABITATS.—Woods, lanes, hedgebanks, streamsides. It grows with greatest luxuriance in the shade, and in positions where accumulations of leaf-mould have been formed. Small specimens may sometimes be found on rocks and even on old walls, but these are not the natural habitats of this species, which requires a depth of rich earth and a sloping position to acquire its finest proportions.

WHERE FOUND.—In *England*, in the counties of Berks, Bucks, Cambridge, Chester, Cornwall, Cumberland, Derby, Devon, Dorset, Durham, Essex, Gloucester, Hants (the mainland and the Isle of Wight), Hereford, Hertford, Kent, Lancaster, Leicester, Lincoln, Middlesex, Monmouth, Norfolk, Northampton, Northumberland, Nottingham, Oxford, Salop, Somerset, Stafford, Suffolk, Surrey, Sussex, Warwick, Westmoreland, Wilts,

Worcester, and York. In *Wales*, in the counties of Anglesea, Brecknock, Caermarthen, Caernarvon, Cardigan, Denbigh, Flint, Glamorgan, Merioneth, Pembroke, and Radnor. In *Scotland*, in the counties of Aberdeen, Argle, Ayr, Banff, Berwick, Caithness, Clackmannan, Dumbarton, Edinburgh, Elgin, Fife, Forfar, Haddington, Inverness, Kincardine, Kinross, Lanark, Linlithgow, Orkney, Perth, Renfrew, Ross, Roxburgh, Stirling, and Sutherland; also in the islands of Arran, Cantyre, Harris, Islay, Lewis, and Uist. In *Ireland*, in the counties of Clare, Cork, Down, Dublin, Galway, and Kilkenny; in King's County, Limerick, Tipperary, Waterford, and Wicklow. In Jersey and Guernsey. It grows from the sea-level to three thousand seven hundred feet above it.

———◆◇◆———

XXXII.—THE HAY-SCENTED BUCKLER FERN.

Lastrea recurva.

(Plate VI., Fig. 2, page 59.)

LENGTH OF FROND.—One foot to two feet.

GENERAL DESCRIPTION.—*Roots* abundant, wiry, fibrous. *Rootstock*, a tufted cormus, whose crown is slightly raised above the soil. *Fronds* strongly resembling in general form—except in the matter of size—those of *Lastrea dilatata*. Stipes varying in length, but frequently about as long as the leafy part, scattered over near its base, and also in a less degree higher up, with a few dark or muddy-brown scales; leafy part triangular, tripinnate in its lower part and bipinnate above; pinnæ opposite or alternate, lower ones also triangular, succeeding ones above becoming narrower and narrower towards the apex of the frond; pinnules alternate on the secondary rachides, and more or less deeply divided into sharply-indented lobes, the pinnules on the under sides of the midstems of the lowest pair of pinnæ being considerably longer and more divided than those on the upper sides of the same midstems; the same kind of difference, though in a less degree, being observable in the pinnules of the pinnæ above—the difference gradually disappearing towards the apex of the frond.   Characters which, besides its smaller size, distinguish this species from *Lastrea dilatata* are the strong hay scent which is diffused by the fronds, especially when in a dry or drying state, the bluish-green hue of its fronds, and the recurving of the lobes of the pinnules.   It has been seen that in *Lastrea dilatata* the under sides of the lobes are concave, a feature which gives a drooping aspect to the entire frond.   In *Lastrea recurva*, on the other hand, the lobes are recurved, so that they are slightly concave on their upper sides.   *Fructification* distributed over the entire under surface of the frond, and consisting of rows—one on each side of the midvein of each lobe in the lower part of the frond and of each pinnule on the upper part—of kidney-shaped indusia, green at first, and subsequently becoming brown and falling off as the spores are ripened.

HABITATS.—Moist and sheltered rocky and other banks and hollows of woods ; loose stones upon hillsides or embankments ; the tops and sides of hedgebanks where the luxuriance of shrubs and trees makes shady places.   The positions this fern prefers are those where rich leaf-soil is found in conjunction with shade and moisture.

WHERE FOUND.—In *England*, in the counties of Cornwall, Cumberland, Devon, Dorset, Hereford, Kent, Lancaster, Northumberland, Salop, Somerset, Sussex, Wéstmoreland, and York.   In *Wales*, in the counties of Anglesea, Caernarvon, Glamorgan, Merioneth and Pembroke.   In the Isle of Man.   In *Scotland*, in the counties of Argyle, Berwick, Dumbarton, Forfar, Inverness, Orkney, and Roxburgh; also in the islands of Arran, Mull, and North Uist. In *Ireland*, in the counties of Antrim, Clare, Cork, Donegal, Galway, Kerry, Londonderry, Mayo, Sligo, Waterford, and Wicklow. In the island of Guernsey. It is found growing from the sea-level to two or three thousand feet above it.

XXXIII.—THE RIGID BUCKLER FERN.

Lastrea rigida.

(Plate VI., Fig. 3, page 59.)

LENGTH OF FROND.—One foot to two feet.
GENERAL DESCRIPTION. — *Roots* abundant, fibrous. *Rootstock*, a thick, tufted cormus. *Fronds* rigid, erect; stipes about half the length of, or as long as, the leafy part, scaly, the scales being continued along the rachis; leafy part triangular, bipinnate, pinnæ cone-shaped, in pairs or alternate upon the rachis, and divided into oblong, alternate, indented pinnules, which are arranged in symmetrical order upon the secondary rachides— the whole frond having a very elegantly-cut appearance. *Fructification* consisting of lines of sporangia—a line on each side of the midvein of each pinnule—covered by the kidney-shaped indusia characteristic of the genus *Lastrea.*
HABITATS.—Rocky hollows, in the moist crevices of which the Rigid Buckler Fern often grows abundantly in its own districts. It prefers limestone rocks; and is, in fact, the only one of the genus *Lastrea* which prefers rocky habitats.
WHERE FOUND.—In *England*, only in the counties of Cornwall, Lancaster, Somerset, Westmoreland, and York;

I

in Lancashire, in the neighbourhood of Silverdale; in Westmoreland, at Arnside Knot, at Farlton Knot, and at Hutton Roof Crags; in Yorkshire, at Ingleborough, Ingleton, on the Attermine Rocks, near Settle, at Wharnside, and White Scars. It has not been recorded as having been found in *Wales* or in *Scotland*, and in *Ireland* only in the county of Louth. It is found growing at various heights up to fifteen hundred feet above the sea-level.

———◆◇◆———

XXXIV.—THE CRESTED BUCKLER FERN.

Lastrea cristata.

(Plate XIV., Fig. 1, page 75.)

LENGTH OF FROND.—One to three feet.

GENERAL DESCRIPTION. — *Roots* abundant, fibrous. *Rootstock*, a stout caudex, which extends itself laterally in the ground, producing several crowns, which oftentimes, when the plant spreads over an area of several square feet, are still adherent to each other, and show their common origin. *Fronds* numerous, produced promiscuously from the crowns without any particular order, such as that noticed in the shuttlecock-shapes of the sets of fronds of several other species of the same genus; stipes brittle, rather short—not exceeding usually one-half the length of the leafy part—and having a few light-brown scales scattered upon it; leafy part, narrowly triangular, or lanceolate, nearly, but not quite, bipinnate; pinnæ opposite or alternate upon the rachis, triangular, pinnatifid, divided, nearly down to their midstems, into oblong, indented pinnules, which are attached to the secondary rachides by the whole width of their bases. The habit of the frond is very erect, and the arrangement of pinnæ and pinnules very symmetrical. *Fructification* produced over the whole under sides of the fronds, and consisting of rows of sori, one on each side of the midvein of each pinnule—each sorus covered by a kidney-shaped indusium in the earlier stage of growth. The indusia, however, fall away and disappear on the ripening of the spores.

HABITATS.—Shady, boggy places, oftentimes under shrubs or trees in such situations. Though the habitats of this species are thus marshy, it is invariably found to prefer little mounds, knolls, or other elevations a few inches above the surface of the bog. Bog tree-stumps upon which have accumulated leaf-soil, grass, and moss, are amongst the favourite places for the finding of *Lastrea cristata*, which, however, is local in its appearance, and not widely distributed.

WHERE FOUND.—In *England*, in the counties of Chester, Huntingdon, Norfolk, Nottingham, Suffolk, and York. In Cheshire, it has been found in the Wybunbury Bog; in Norfolk, at Bawsey Heath, near Lynn, near Dersingham, between Hunstanton and Lynn ; at Edgefield, near Holt; at Fritton, near Yarmouth ; and at Surlingham Broad, near Norwich ; in Nottinghamshire, on the Bulwell Marshes and in Oxton Bogs (although it may possibly at the present time have become extinct on the Bulwell Marshes) ; in Staffordshire, near Madeley, and in a bog in the vicinity of Newcastle-under-Lyne ; in Suffolk, at Bexley Decoy, near Ipswich, and at Westleton ; in Yorkshire, near Knaresborough and near Malton. It is said to have been found in *Scotland* only in Renfrew and Wigtonshire, and neither in *Wales* nor *Ireland*. It grows generally at low elevations not exceeding three hundred feet above the sea-level.

XXXV.—THE PRICKLY-TOOTHED BUCKLER FERN.

Lastrea spinulosa.

(Plate VII., Fig. 2, page 61.)

LENGTH OF FROND.—One foot to three feet.
GENERAL DESCRIPTION.—*Roots* abundant, fibrous.
Rootstock, a tufted caudex, which extends into numerous
crowns that are noticeable by the absence of scales.
Fronds numerous, triangular, deciduous, bipinnate, sometimes, in the lower part of the frond, nearly tripinnate ;
pinnæ more or less triangular, opposite or alternate on
the rachis, and divided into oblong, sharply-incised pinnules, furnished with spinous, bristle-like points which
are turned towards the apices of the pinnules. As in
the case of *Lastrea dilatata* and *Lastrea recurva,* the lower
pairs of pinnæ are more developed than the upper ones,
the basal pinnules of these being elongated, and again
divided into spiny lobes. The pinnæ—especially the
lower pairs—are usually pointed upwards in a direction
diagonal to that of the rachis. The stipes is generally
about the same length as the leafy part of the frond,
though sometimes longer, and is brittle, and furnished
near the base with a few light-brown scales. *Fructification* produced in rows of small sori, covered by kidney-
shaped indusia, and scattered equally over the under
sides of the fronds—a row of sori on each side of the
midvein of each pinnule or lobe according to its size
and position.

HABITATS.—Similar in all respects to those of *Lastrea
cristata*—namely, boggy places of low-lying heaths and
moorlands, especially in places where, under the shelter
of shrub or tree, little grassy or mossy knolls have been

formed above the general bog or marsh level. When
the boggy soil is of peat and leaf-mould the most favourable conditions of growth are provided for this species.

WHERE FOUND.—In *England*, in the counties of Bed-
ford, Berks, Bucks, Cambridge, Chester, Cornwall,
Cumberland, Derby, Devon, Dorset, Durham, Essex,
Gloucester, Hants (the mainland and the Isle of Wight),
Hereford, Hertford, Huntingdon, Kent, Lancaster,
Leicester, Lincoln, Middlesex, Monmouth, Norfolk,
Northampton, Northumberland, Nottingham, Oxford,
Salop, Somerset, Stafford, Suffolk, Surrey, Sussex,
Warwick, Westmoreland, Worcester, and York. In
Wales, in the counties of Anglesea, Brecknock,
Caermarthen, Caernarvon, Flint, Glamorgan, and Merio-
neth. In the Isle of Man. In *Scotland*, in the counties
of Aberdeen, Argyle, Dumbarton, Dumfries, Edinburgh,
Elgin, Fife, Forfar, Inverness, Kincardine, Kinross,
Perth, Renfrew, Ross, Roxburgh, and Stirling. Also in
the isles of Harris, Lewis, and Uist. Its range upwards
from the sea-level does not extend beyond some three
hundred feet.

XXXVI.—THE MOUNTAIN BUCKLER FERN.

Lastrea montana.

(Plate VII., Fig. 1, page 61.)

LENGTH OF FROND.—One foot to four feet and a half.
GENERAL DESCRIPTION.—*Roots* abundant, long, wiry,
fibrous. *Rootstock*, a short, stout, tufted cormus, whose
crown is raised slightly above the surface of the ground.
Fronds deciduous, abundant, lemon-scented, erect-grow-
ing, produced in an arrangement shuttlecock-shape
around the crown, which is furnished with silvery-looking
scales in place of the rust-coloured scales on the crown
of the Male Fern, a species which *Lastrea montana* very

much resembles in some other respects ; stipes very
short, straw-coloured—as is also the rachis—and furnished with a few, light-coloured scales, which are often
continued upon, and a short way along, the rachis ; leafy
part lance-shaped, widest about the middle, pointed at
the apex, and tapering gradually at the base until the
pinnæ are less than half an inch long ; pinnæ opposite
or alternate upon the rachis, long, narrow, pointed, widest
at the base—each pinna pinnatifid and more or less
deeply cleft into oblong, blunt-pointed pinnules. *Fructification* marginal, produced in lines of sori along the
two margins of each pinnule, most abundant on the
upper side of the frond ; sori partially indusiate, the
indusia consisting of little rounded scales situated upon
the centre of the sori, and soon falling off as the period
of spore-ripening arrives.

HABITATS.—Open heaths ; moors ; the more open
parts of woods and forests ; hillsides ; mountain-sides ;
streamsides. In many cases it completely occupies the
ground. On the ground between stones that border
moorland streams this species may often be seen growing in great beauty and luxuriance. Its presence is
ordinarily very conspicuous—its golden-green fronds
covering hillsides with their wealth of golden green, and
perfuming the air with their balsamic fragrance.

WHERE FOUND.—In *England,* in the counties of Bucks,
Chester, Cornwall, Cumberland, Derby, Devon, Dorset,
Durham, Essex, Gloucester, Hants (the mainland and
the Isle of Wight), Hereford, Hertford, Kent, Lancaster,
Leicester, Lincoln, Middlesex, Monmouth, Norfolk,
Northampton, Northumberland, Nottingham, Oxford,
Rutland, Salop, Somerset, Stafford, Suffolk, Surrey,
Sussex, Warwick, Westmoreland, Wilts, Worcester, and
York. In *Wales,* in the counties of Anglesea, Breck-
nock, Caermarthen, Caernarvon, Cardigan, Denbigh,
Flint, Glamorgan, Merioneth, Pembroke, and Radnor.
In the Isle of Man. In *Scotland,* in the counties of

Aberdeen, Argyle, Ayr, Berwick, Caithness, Clack-
mannan, Dumbarton, Dumfries, Edinburgh, Elgin, Fife,
Forfar, Inverness, Kincardine, Kinross, Lanark, Perth,
Renfrew, Ross, Roxburgh, Stirling, and Sutherland;
also in the isles of Arran, Cantyre, Islay, Shetland, and
Uist. In *Ireland*, in the counties of Clare, Donegal,
Galway, Kerry, Londonderry, Waterford, and Wicklow.
It is found growing at various altitudes up to three
thousand feet above the sea-level.

XXXVII.—THE MARSH BUCKLER FERN.

Lastrea thelypteris.

(Plate X., Fig. 1, page 67.)

LENGTH OF FROND.—Barren fronds, one foot to three
feet; fertile fronds, a foot to four feet.

GENERAL DESCRIPTION.—*Roots* black, fibrous, abundant. *Rootstock*, an extensively-creeping rhizoma, slender
and blackish. Fronds of two kinds—barren and fertile
—numerous, light green, herbaceous, brittle, fragile;
stipes about equal to the leafy part, very thin, pale green,
delicate, and brittle; leafy part lance-shaped, broadest in
the centre, tapering to a somewhat blunt point at the
apex, and tapering slightly towards the base; pinnæ
opposite or alternate, and somewhat distant along the
rachis, long, narrow, pointed, broadest at the base,
pinnatifid—each pinna deeply cleft into thin, plain,
oblong, entire pinnules. The pinnules of the fertile
fronds, besides being longer, are somewhat more contracted than those of the barren ones. *Fructification*
borne in rows of sori upon the under sides of the pinnules,
midway between their midveins and their margins, each
sorus roundish in shape and covered by a roundish
indusium, which, however, soon falls off and disappears.

HABITATS.—Wet marshes and liquid bogs. It is especially luxuriant in positions where shade and shelter are provided by shrubs or trees. No other British fern selects habitats which are so absolutely watery as are those favoured by the Marsh Buckler Fern, which grows actually in the soft liquid ooze of bogs, its rhizomas *floating* on the bog surfaces.

WHERE FOUND.—In *England*, in the counties of Bedford, Berks, Cambridge, Chester, Cumberland, Devon, Dorset, Essex, Hants (the mainland and the Isle of Wight), Hereford, Huntingdon, Kent, Leicester, Lincoln, Norfolk, Northumberland, Nottingham, Salop, Somerset, Stafford, Suffolk, Surrey, Sussex, Warwick, Westmoreland, and York. In *Wales*, in the counties of Anglesea, Caernarvon, Flint, Glamorgan, and Pembroke. In *Scotland*, only in the county of Forfar. In *Ireland*, in the counties of Antrim, Galway, Kerry, and Mayo.

XXXVIII.—THE FORKED SPLEENWORT.

Asplenium septentrionale.

(Plate XIV., Figs. 8 and 9, page 75.)

LENGTH OF FROND.—Two to six inches.

GENERAL DESCRIPTION.—*Roots* long, wiry, very fine, abundant, fibrous. *Rootstock* very small, tufted. *Fronds* numerous, evergreen, grass-like, usually produced in dense tufts from the crown ; stipes pale green, purplish-brown at the base, three or four times longer than the leafy part, which consists of two or three narrow, simple, or forked branches resembling short blades of grass, each branch being either simple or once or twice sharply cleft at its apex. *Fructification* borne in elongated lines at the backs of the widest leafy part of the frond, the sori distinct and elongated, and covered when young by elongated indusia, but when these fall off becoming confluent upon nearly the whole under side of the frond, and turning then to a dark-brown colour.

HABITATS.—Moist and shady rocky crevices; old walls in positions sheltered by projecting pieces of stone or rock; dark, moist, shady holes or recesses in walls or rocks—hence, generally, this species is inconspicuous, and requires to be carefully sought for.

WHERE FOUND.—In *England*, in Cornwall, Cumberland, Devon, Northumberland, Somerset, Westmoreland, and York; in Cornwall, near Trengwainton Cairn (F. T. Richards); in Cumberland, Borrowdale, Helvellyn, Honister Crags, Keswick, Vale of Newlands, Patterdale, Scawfell, and Wastwater; in Devonshire, on Exmoor; in Northumberland, in crevices of basaltic rocks of Kyloe Crags; in Somersetshire, near the little village of Culbone; in the county of Westmoreland, at Ambleside; and in Yorkshire, upon the rocks of Ingleborough. In *Wales*, in the counties of Caernarvon, Denbigh, and Merioneth. In the county of Caernarvon the habitats of *Asplenium septentrionale* are in the following places: Bettwys-y-Coed, Capel Curig, Carnedd Llewellyn, Craig Dhu, Pass of Llanberis, Llyn-y-cwm, Moel Lechog, and Pont-y-Pair; in the county of Denbigh, rocks at Llan Dethyla in the neighbourhood of Llanrwst. In Merioneth, at Dolgelly (F. T. Richards). In *Scotland*, Aberdeen, Edinburgh, Perth, and Roxburgh; in Aberdeen, on rocks at the Pass of Ballater; in the county of Edinburgh, on rocks at Arthur's Seat, at Blackford Hill, and on other rocks in the same neighbourhood; in Perthshire, in the vicinity of Dunkeld; and in Roxburghshire, at Jedburgh and on the Minto Crags. No habitats of this species have been recorded in *Ireland.* It grows at various altitudes up to three thousand feet above the sea-level.

XXXIX.—THE ALTERNATE SPLEENWORT.

Asplenium germanicum.

(Plate XIV., Figs. 6 and 7, page 75.)

LENGTH OF FROND.—Two to six inches.

GENERAL DESCRIPTION.—*Roots* fibrous, wiry, abundant. *Rootstock* small, tufted. *Fronds* numerous, evergreen, produced in clusters from the crown ; stipes pale-green, purplish-brown at the base, about equal in length to the leafy part ; smooth ; leafy part simply pinnate, with wedge-shaped pinnæ sharply cleft on their upper and broader sides, and placed in alternation on opposite sides of the rachis to which they are attached by short, narrow stems, which broaden and are merged, almost insensibly, into the wider, leafy part of the pinnæ. *Fructification* borne upon the under sides of the wedge-shaped, leafy parts of the pinnæ in elongated—or "linear," as they are called—sori, which run in parallel directions towards the terminal points of the pinnæ. Each elongated sorus is covered when young by a long, green indusium, and is then distinct. But when the indusia are ruptured by the expansion, at ripening, of the sporangia, they burst and are thrown off, and the sori become confluent, covering almost the entire under sides of the pinnæ with a mass of rich, dark-brown spore-cases.

HABITATS.—Rocky crevices similar to those in which *Asplenium septentrionale* grows. The two species are often found growing together.

WHERE FOUND.—In *England*, only in the counties of Cumberland, Northumberland, and Somerset ; in Cumberland, rocks at Borrowdale and on Helvellyn ; in Northumberland, on the Kyloe basaltic rocks ; and in

Somersetshire, at Culbone. In *Wales*, in Caernarvon and Merioneth ; in the former, rocks between Capel Curig and Llanrwst, and rocks at the Pass of Llanberis ; in Merioneth, on Cader Idris (F. T. Richards). In *Scotland*, Edinburgh, Fife, Perth, and Roxburgh, and in the following localities : in Edinburgh, rocks within two miles of the capital ; in the county of Fife, rocks in the neighbourhood of Dunfermline ; in Perthshire, the Stenton Rocks in the neighbourhood of Dunkeld ; in the county of Roxburgh, Minto Crags in the vicinity of Hassendean and rocks on the Tweed near Kelso. It has never been reported from *Ireland*. *Asplenium germanicum* grows at elevations above the sea-level extending from some three hundred to three thousand feet.

XL.—THE RUE-LEAVED SPLEENWORT.

Asplenium ruta-muraria.

(Plate XIII., Figs. 8 and 9, page 73.)

LENGTH OF FROND.—One inch to six inches.

GENERAL DESCRIPTION.—*Roots* fine, wiry, fibrous, and very abundant, growing oftentimes in a dense mass. *Rootstock* short, thick, compact, tufted. *Fronds* evergreen, leathery, dark-green, shining, numerous, produced sometimes in thick tufts from the crown, which is always elevated a little above the surface of the rock or earthy seam of soil upon which it is growing ; stipes smooth, green, purplish-black at the base, equal in length to, or double the length of, the leafy part, or intermediate between these lengths ; leafy part more or less triangular, bipinnate ; pinnæ stalked, alternate upon the rachis and divided, usually, into three wedge-shaped, egg-shaped, or diamond-shaped pinnules, which, in luxuriant specimens are sometimes deeply cleft into unequally-shaped lobes, and where the pinnules are not thus divided their upper and broader edges are more or less conspicuously indented. *Fructification* borne in elongated sori, covered, when young, by pale-green indusia. When they have become disrupted and thrown off the sori, by the enlargement of the sporangia, become confluent and cover the entire under surface of the fronds, turning them to a rich reddish-brown.

HABITATS.—Rocks, brick and stone walls, bridge-arches and old masonry, in shaded positions; but this fern often grows hardily in the sunshine. The parts of walls and rocks selected by these little ferns are generally those where there are more or less moist seams of earth or old crumbling mortar, and it will always be found that the most luxuriant specimens of the Wall Rue are those whose crowns are protected by some jutting portion of stone over them. When the crowns of this little plant are immersed in rocky crevices, so that, though not buried in the earthy seams, the moisture and shade of the crevices surround and protect them from the desiccating effects of sun and wind, they are in the most favourable position for developing luxuriant fronds.

WHERE FOUND.—In *England*, in the counties of Bedford, Berks, Bucks, Cambridge, Chester, Cornwall, Cumberland, Derby, Devon, Dorset, Durham, Essex, Gloucester, Hants (the mainland and the Isle of Wight), Hereford, Hertford, Kent, Lancaster, Leicester, Lincoln, Middlesex, Monmouth, Norfolk, Northampton, Northumberland, Nottingham, Oxford, Rutland, Salop, Somerset, Stafford, Suffolk, Surrey, Sussex, Warwick, Westmoreland, Wilts, Worcester, and York. In *Wales*, in the counties of Anglesea, Cardigan, Caermarthen, Caernarvon, Denbigh, Flint, Glamorgan, Merioneth, Montgomery, Pembroke, and Radnor. In *Scotland*, in the counties of Aberdeen, Argyle, Ayr, Banff, Berwick, Caithness, Clackmannan, Cromarty, Dumbarton, Dumfries, Edinburgh, Elgin, Fife, Forfar, Haddington, Inverness, Kincardine, Kinross, Kirkcudbright, Lanark, Linlithgow, Nairn, Orkney, Peebles, Perth, Renfrew, Roxburgh, Selkirk, Stirling, and Sutherland; also in the islands of Ailsa Craig, Cantyre, Harris, Iona, Islay, and Uist. In *Ireland*, in the counties of Armagh, Clare, Cork, Down, Dublin, Galway, Kerry, and Kilkenny; in King's County, Limerick, Louth, Tipperary, Waterford, and Wicklow. It is also found in Jersey. *Asplenium ruta-muraria* grows at elevations extending to about two thousand feet above the sea-level.

---◦◇◦---

XLI.—THE BLACK MAIDENHAIR SPLEENWORT.

Asplenium adiantum-nigrum.

(Plate III., page 53.)

LENGTH OF FROND.—An inch to two feet, according to its more or less exposed, sunny and stony, or sheltered, shady, moist, and in other ways congenial position.

GENERAL DESCRIPTION.—*Roots* long, fibrous, wiry, abundant. *Rootstock* small, tufted, scaly. *Fronds* evergreen, numerous; stipes and rachis more or less purple; stipes equal in length to the leafy part, sometimes a little shorter, and sometimes a little longer; leafy part triangular, dark shining-green, with alternated, triangular pinnæ, divided into narrow, elongated, and variously-shaped pinnules, which, in turn, are sub-divided into more or less deeply-indented lobes—the ultimate divisions depending upon the more or less luxuriant state of the plant. *Fructification* produced in the form of elongated sori covered by elongated, pale-green indusia. When

these fall off, the sori, become confluent and densely
cover the whole under side of the frond.

HABITATS.—Walls of all kinds, more or less old, brick
and stone ; ruins, bridge-arches, garden and house walls,
and, indeed, every description of masonry—the luxuriance of the plants depending upon the greater or less
accumulation of leaf-mould in the moist holes or seams
of soil in rock or wall, and upon the greater or less
amount of shade or moisture of the position. Stony
banks, or soil covered by large or small pieces of stone,
such as hedgebanks, streambanks, or the banks formed
by cuttings through hilly, rocky, or moorland country,
are also the favoured habitats of this beautiful species.
Where, on such banks, shrubs, growing from between
the stones, give shelter, and, at the same time, provide
—by the annual deposit of leaves—for the enrichment
of the soil, *Asplenium adiantum-nigrum* grows in its
finest form.

WHERE FOUND.—In *England*, in the counties of
Bedford, Berks, Bucks, Cambridge, Chester, Cornwall,
Cumberland, Derby, Devon, Dorset, Durham, Essex,
Gloucester, Hants (the mainland and the Isle of Wight),
Hereford, Hertford, Kent, Lancaster, Leicester, Lincoln,
Middlesex, Monmouth, Norfolk, Northampton, Northum-
berland, Nottingham, Oxford, Rutland, Salop, Somerset,
Stafford, Suffolk, Surrey, Sussex, Warwick, Westmore-
land, Wilts, Worcester, and York. In *Wales*, in the
counties of Anglesea, Caermarthen, Caernarvon, Car-
digan, Denbigh, Flint, Glamorgan, Merioneth, Mont-
gomery, Pembroke, and Radnor. In the Isle of Man.
In *Scotland*, in the counties of Aberdeen, Argyle, Ayr,
Banff, Berwick, Caithness, Clackmannan, Cromarty,
Dumbarton, Dumfries, Edinburgh, Elgin, Fife, Forfar,
Haddington, Inverness, Kincardine, Kinross, Kirkcudbright, Lanark, Linlithgow, Nairn, Orkney, Perth, Renfrew, Ross, Roxburgh, Selkirk, and Sutherland ; also
in the islands of Ailsa Craig, Arran, Cantyre, Harris,

Iona, and Islay. In *Ireland*, in the counties of Antrim, Clare, Cork, Down, Dublin, Galway, Kerry, and Kilkenny; in King's County, Limerick, Louth, Meath, Tipperary, Waterford, and Wicklow ; also in the Arran Isles. It is found growing at various elevations extending up to nearly two thousand feet about the sea-level.

———◦◇◦———

XLII.—THE LANCEOLATE SPLEENWORT.

Asplenium lanceolatum.

(Plate XIII., Figs. 2 and 3, page 73.)

LENGTH OF FROND.—Four to eighteen inches.
GENERAL DESCRIPTION.—*Roots* long, fibrous, wiry, abundant. *Rootstock* somewhat large, dark brown, scaly, tufted. *Fronds* evergreen, lance-shaped (distinguished by this feature from the triangular fronds of *Asplenium adiantum-nigrum*, which it otherwise resembles) ; stipes a third the length of the leafy part and sometimes less in proportion, purplish red in colour, the same hue being noticeable, in a greater or less degree, on the rachis ; leafy part bright green, bipinnate ; pinnæ opposite ‐or alternate on the rachis, narrowly triangular, divided into alternate and—in well-developed specimens —distinctly stalked, fan-shaped, or four-sided and indented pinnules. *Fructification* produced over the entire under surface of the frond, and consisting of sori which, though elongated—as in the Spleenworts generally—are less elongated than those of *Asplenium adiantum-nigrum*. When the indusia fall off, the sori become rounded in form and somewhat bulged out as the sporangia increase by development ; but each sorus ordinarily remains distinct from the others, and thus presents another feature which distinguishes this species

from *Asplenium adiantum-nigrum*, the sori in which ordinarily become confluent.

HABITATS.—Shady positions on or near the sea-coast; moist and dripping rocks; the shady sides of cliffs; sea-caverns; rocky holes, oftentimes almost dark. This species is especially luxuriant in places where water oozes or trickles over the face of cliff or other rock, or along the internal sides of caverns, crevices, or other holes or fissures of rocks. Soft rock seamed with vegetable mould offers, where the aspect and atmosphere are congenial, especially favourite habitats for the Lanceolate Spleenwort.

WHERE FOUND.—In *England*, in the counties of Cornwall, Devon, Gloucester, Kent, Somerset, Sussex, and Yorkshire. Amongst its habitats in Cornwall are sea-rocks, or rocks adjacent to the sea-coast, at Land's End, Penzance, and St. Ives. In Devonshire, along the rivers Dart, Plym, Tamar, and Tavy, especially near and at the mouths of those rivers. On the south-eastern sea-coast of Devon, especially from Portlemouth to Prawle Point and at Salcombe. The Yorkshire habitat of *Asplenium lanceolatum* is a newly-found one, and its discovery was first communicated to the author of this volume by the Rev. R. Gatty, of Bradfield Rectory, who kindly furnished fronds from the specimens he had taken in this northern county. In *Wales*, in the counties of Caernarvon, Denbigh, Glamorgan, Merioneth, and Pembroke. No habitats of this species have been discovered in *Scotland*, and only one in *Ireland*, namely, near the town of Cork. It is, however, abundant in Jersey, in Guernsey, and in Sark.

XLIII.—THE ROCK SPLEENWORT.

Asplenium fontanum.

(Plate XIII., Figs. 6 and 7, page 73.)

LENGTH OF FROND.—Three inches to a foot, the maximum length given being, however, exceptional.
GENERAL DESCRIPTION.—*Roots* fibrous, wiry, abundant. *Rootstock* small, tufted, erect. *Fronds* numerous, stiff, evergreen, narrowly lance-shaped ; stipes purplish-black, very short, the leafy pinnæ being continued almost close to the crown, leaving oftentimes no more than half an inch of clear stipes ; leafy part pinnate, light green, pointed at the apex, broadest near its centre, and diminishing downwards ; pinnæ opposite or alternate upon the rachis, very short, either triangular or egg-shaped, and either sharply indented or—in large specimens— again divided into somewhat four-sided, indented pinnules. *Fructification* produced in sori which are slightly oblong, and are covered by slightly-oblong in-dusia. When these fall off, the sori frequently become rounder and confluent, though they are perhaps as frequently distinct from each other.
HABITATS.—Moist, shady fissures of rocks, and crevices of walls ; sea-cliffs and sea-caves. This species grows under very much the same conditions, and in very much the same positions, as *Asplenium lanceolatum.*
WHERE FOUND.—In *England,* only in the counties of Derby, Dorset, Hants, Northumberland, Westmoreland, and York. In Derbyshire, near Matlock ; in Dorset-shire, in the Swanage Cave, Isle of Purbeck ; in Hamp-shire, near Petersfield ; in Northumberland, near Alnwick Castle ; in Westmoreland, near Wybourn. In York-

K

shire, in Wharncliffe Wood. In *Wales*, only in the county of Caernarvon, between Tan-y-Bwlch and Tremadoc. In *Scotland*, only in Kincardineshire, on rocks in the vicinity of Stonehaven. In *Ireland*, it has been found at Cavehill, near Belfast.

XLIV.—THE GREEN SPLEENWORT.

Asplenium viride.

(Plate XIV., Figs 4 and 5, page 75.)

LENGTH OF FROND.—Two to ten inches.

GENERAL DESCRIPTION.—*Roots* fibrous, wiry, abundant, *Rootstock* small, tufted. *Fronds* numerous, evergreen, produced in tufts from the crown, narrow, tapering, broadest about the centre, simply pinnate; stipes rather short, green, purplish at the base; rachis also green; pinnæ light green, opposite or alternate, attached to the rachis by very short but distinct stalks, roundish-oblong in shape, finely indented upon their margins, usually largest about the centre, diminishing in size towards the apex and towards the base of the frond. This species bears a strong general resemblance to its much more common and widely-distributed congener *Asplenium trichomanes ;* but the especial and immediate mark of distinction lies in the colour of the stipes and of the lower part of the rachis, a colour which in *Asplenium viride* is always green—except at the base of the stipes— and in *Asplenium trichomanes* always purple. *Fructification* produced in narrow, oblong sori, covered by indusia, and occupying nearly the centre of each little pinna, becoming confluent about the centre of the pinnæ when the indusia have fallen away, and not spreading, as is usually the case with the fructification of the Common Maidenhair Spleenwort, on the entire leafy under sides.

HABITATS.—Wild outlying districts, away, ordinarily, from the immediate vicinity of towns; rocks where trickling moisture can flow over the crowns of these little plants. The most moist and shady of rocky crevices are the favoured habitats.

WHERE FOUND.—In *England*, in the counties of Chester, Cumberland, Derby, Durham, Lancaster, Leicester, Middlesex, Monmouth, Northumberland, Stafford, Surrey, Sussex, Westmoreland, Worcester, and York. The particular localities in these counties are the following, which will be mentioned in the alphabetical order, first of the counties and then of the districts, in or near which this species is found. In Cheshire, Carr-edge; in the county of Cumberland, Ashness Gill, Borrowdale, Borrow Force (a "force" is the north-country name of a waterfall), and Gillsland; in Derbyshire, Buxton, Castleton, Cavedale, and Dovedale; in Durham, Falcon Clints, Teesdale, and Weardale; in Kent, Maidstone; in the county of Lancaster, Dulesgate and Staley; in Leicestershire, Beacon Hill and Charley Forest; in Middlesex, Southgate; in the county of Northumberland, banks of the River Irthing; in Staffordshire, Dovedale; in the county of Surrey, Mickleham; in Sussex, Danny; in Westmoreland, Ambleside, Arnside, Casterton Fell, Farlton, Hutton Roof, Kendal Fell, Mazebec Scar, and Patterdale; in Worcestershire, Ham Bridge; in Yorkshire, Aix-la-Beck, Craven, Gordale, Leeds, Ogden Clough (in the neighbourhood of Halifax), Ingleborough, Reeth Moor, Richmond, Settle, Swaledale, Wensleydale, and Widdal Fell. In *Wales*, in the counties of Brecknock, Caernarvon, Glamorgan, and Merioneth; and in the following localities in those counties: in Brecknockshire, Brecon Beacon, and Trecastle Beacon (Brecon), Capel Colbren, and Capel-y-Fin; in the county of Caer-

narvon, Clogwyn-y-Garnedd, Clogwyn-du-Yrarddu, Cwm
Idwal, Glyder Vawr, Glyn-y-Cwm, and Twll-du ; in the
county of Glamorgan, Cilhepste Waterfall (Pont Nedd
Vechan), Darran-yr-Ogof, Merthyr Tydfil, and Ystrad-
gunlais ; in the county of Merioneth, Cader Idris.   In
Scotland, in the counties of Aberdeen, Argyle, Ayr,
Clackmannan, Dumbarton, Dumfries, Edinburgh, Elgin,
Fife, Forfar, Inverness, Kinross, Lanark, Linlithgow,
Nairn, Perth, Ross, Stirling, and Sutherland.   The
following are some of the localities in those counties :
in Argyleshire, Dunoon ; in the county of Dumfries,
Mare's-tail ; in Forfarshire, Canlochen, Clova ; in Lan-
arkshire, falls of the Clyde ; in the county of Nairn,
Cawdor Woods ; in Perthshire, Ben Chonzie (Crieff),
Ben Lawers, Ben Voirlich, Blair Athol, and Drummond
Hill ; in Sutherlandshire, Assynt ; also in the Shetland
Isles and the Isle of Mull.   In *Ireland*, in the counties
of Cork, Donegal, Killarney, Kerry, and Sligo.   Subjoined are the parts of those counties : in Cork, Bandon ;
in Donegal, Lough Eask ; in Kerry, Tork Mountains ;
and in Sligo, Ben Bulben.   It occurs at various altitudes up to two thousand five hundred feet above the
sea-level.

XLV.—THE COMMON MAIDENHAIR SPLEENWORT.

Asplenium trichomanes.

(Plate XIV., Figs. 2 and 3, page 75.)

LENGTH OF FROND.—Two to eighteen inches, the
maximum length being very exceptional.

GENERAL DESCRIPTION.—*Roots* fibrous, long, abundant, wiry. *Rootstock* somewhat large for the size of the
plant, tufted. *Fronds* evergreen, produced in numerous

tufts from the crown, tapering, widest about the middle, tapering at each end, simply pinnate, stipes very short, wholly purple ; rachis also purple, in this respect being distinguished from *Asplenium viride*, which is much like it in other respects. Pinnæ deep green, small, oval, entire, opposite or alternate on the rachis, usually in opposite or nearly opposite pairs, seldom reaching a quarter of an inch in length. *Fructification* produced in oblong sori, covered by oblong indusia. When these fall off, the sporangia become confluent over the entire under surfaces of the pinnæ—in this respect also differing from *Asplenium viride*, whose sporangia, when they become confluent, occupy only the centre of the pinnæ, leaving a green, leafy margin around them.

HABITATS.—Rocks, walls, and old masonry of all kinds, especially where, in the crevices which may have been formed, leaf-soil has accumulated and moisture has entered. Hence rocks or stony places by running streams, bridge-arches, stone parapets, dwelling-house and garden-walls, out-buildings, cliffs, and stony embankments of all kinds. On the drier sides of such habitats it is often stunted and puny, whilst on the shady, moist, crumbling surfaces of rock or wall it becomes much larger. In hedge and other embankments, where the surface is sheltered by shrubs and the soil is rich—especially where its roots are snugly ensconced under fragments of stone which may lie upon the face of an incline—this species assumes its finest and most luxuriant proportions. But such fine specimens require, ordinarily, to be sought for, as, hid beneath the friendly shelter of the superincumbent bushes, which promote the shadiness, the moisture, and the richness of soil of their habitats, they are not conspicuous, and are often entirely concealed from the passer-by.

WHERE FOUND.—In *England*, in the counties of Bedford, Berks, Bucks, Cambridge, Chester, Cornwall, Cumberland, Derby, Devon, Dorset, Durham, Essex,

Gloucester, Hants (the mainland and the Isle of Wight), Hereford, Hertford, Kent, Lancaster, Leicester, Lincoln, Middlesex, Monmouth, Norfolk, Northampton, Northumberland, Nottingham, Oxford, Rutland, Salop, Somerset, Stafford, Suffolk, Surrey, Sussex, Warwick, Westmoreland, Wilts, Worcester, and York. In *Wales*, in the counties of Anglesea, Brecknock, Caermarthen, Caernarvon, Denbigh, Flint, Glamorgan, Merioneth, Montgomery, Pembroke, and Radnor. In the Isle of Man. In *Scotland*, in the counties of Aberdeen, Argyle, Ayr, Banff, Berwick, Caithness, Clackmannan, Cromarty, Dumbarton, Dumfries, Edinburgh, Elgin, Fife, Forfar, Haddington, Inverness, Kincardine, Kinross, Kirkcudbright, Lanark, Linlithgow, Nairn, Orkney, Peebles, Perth, Renfrew, Ross, Roxburgh, Selkirk, Stirling, and Sutherland ; also in the islands of Arran, Bute, Cantyre, Islay, and Harris. In *Ireland*, in the counties of Antrim, Clare, Cork, Down, Dublin, Galway, Kerry, and Kilkenny ; in King's County, Limerick, Louth, Tipperary, Tyrone, and Wicklow. In the Channel Islands. It grows at various heights, extending to some two thousand feet above the sea-level.

XLVI.—THE SEA SPLEENWORT.

Asplenium marinum.

(Plate X., Figs. 4 and 5, page 67.)

LENGTH OF FROND.—Two to eighteen inches, the maximum length being exceptional.

GENERAL DESCRIPTION.—*Roots* fibrous, rather fleshy, and abundant. *Rootstock* stout, erect, tufted, with scales upon its crown. *Fronds* evergreen, lance-shaped, leathery, shining, simply pinnate ; stipes smooth, purple, about half the length of the leafy part, and sometimes shorter than that; rachis often purple, sometimes purple on the lower part and green higher up; leafy part widest about the middle, tapering to a blunt point at the apex, and tapering generally, but not always, by the diminution of the pinnæ towards the base; pinnæ in opposite pairs, or alternate upon the rachis, indented, wing-shaped, or ear-shaped, ordinarily attached by their narrow, stalk-like bases to narrow, leafy margins or wings, which run along on each side of the rachis. *Fructification* pro-duced in elongated sori, covered by elongated indusia, and placed diagonally between the midveins and the margins of the pinnæ. Though generally, even when ripened, distinct, the lines of sori become sometimes confluent—turning to a rich brown, which conspicuously contrasts with the deep green of the pinnæ.

HABITATS.—Sea-caverns; cliffs or other rocks in or very near the sea. It is very rarely that this fern is found growing far from the coast, though it not un-frequently is found of a more or less diminutive size upon rocks in tidal rivers several miles from the sea. Its favourite positions are moist and shady crevices of the open sides of cliffs, especially in situations where water oozes through such crevices or trickles down the out-ward face of the rock. Shady clefts, formed by jutting pieces of rock, moist corners at the entrance to cliff hollows or caverns; cavern roofs; rocks detached from the coast and surrounded by the sea. These and the under sides of rocks overhanging the mouths of tidal rivers and similar rocks further inland are, one and all, favoured habitats of *Asplenium marinum.*

WHERE FOUND.—In *England,* on the coasts of the counties of Chester, Cornwall, Cumberland, Devon, Dorset, Durham, Gloucester (banks of the Severn), Hants (the Isle of Wight), Lancaster, Northumberland, Somerset, Sussex, Westmoreland, and York. In *Wales,* on the coasts of the counties of Anglesea, Caermarthen,

Caernarvon, Cardigan, Glamorgan, Merioneth, and
Pembroke. On the coasts of the Isle of Man. In
Scotland, on the coasts of the counties of Aberdeen,
Argyle, Ayr, Banff, Berwick, Caithness, Cromarty, Dum-
barton, Edinburgh, Elgin, Fife, Forfar, Kincardine,
Kinross, Kirkcudbright, Linlithgow, Nairn, Orkney,
Perth, Renfrew, Ross, Stirling, Sutherland, and Wigton.
Also on the coasts of the isles of Ailsa Craig, Cantyre,
Harris, Iona, Islay, Lewis, and Uist. In *Ireland*, on the
coasts of the counties of Clare, Cork, Down, Dublin,
Galway, Kerry, Limerick, Louth, Waterford, and Wick-
low : also on the coasts of the isles of Arran. It is
also found on the coasts of Jersey and Guernsey.

XLVII.—THE SCALY SPLEENWORT.

Asplenium ceterach.

(Plate XIII., Figs. 4 and 5, page 73.)

LENGTH OF FROND.—An inch to eight inches.
GENERAL DESCRIPTION.—*Roots* long, fibrous, wiry,
very abundant, oftentimes forming dense masses. *Rootstock* tufted, scaly. *Fronds* not numerous, thick, leathery,
evergreen, produced in an irregular circle around the
crown ; pinnatifid ; stipes, very short, scaly ; leafy part
lance-shaped, and, though generally pinnatifid, sometimes
in the lower part of the frond partially pinnate—the deep,
wide indentations and the lobes formed by them being
rounded and waved on each side of the rachis in a
manner somewhat similar to that of a large saw. The
upper surface of the leafy part is bluish-green and
velvety to the touch, and the whole under-surface is
densely covered by light reddish-brown or rust-coloured
scales. *Fructification* produced in irregularly-elongated

sori, which are ordinarily quite hidden by the clothing of the scales, and which have imperfect and partially-developed indusia.

HABITATS.—Rocks, old walls, and all kinds of old and crumbling masonry; bridge-arches, house and garden walls, and stony embankments. It grows from the moist, shady seams of its stony habitats, being more or less luxuriant according to the more or less congenial condition of the habitats—leaf-mould in the crevices of rock or wall, caused by the fall and decay of leaves from overarching trees, and a certain amount of moisture, being conducive to vigour and luxuriance. The proof that it is chiefly leaf-mould and not " old mortar "—as is so frequently alleged—that promotes the luxuriant growth of this fern is found in the circumstance that when the walls or rocks on which it is growing are under trees the finest specimens are those amongst loose stones on the tops of such walls or rocks, these being precisely the positions in which there are naturally the largest accumulations of leaf-mould from falling leaves.

WHERE FOUND.—In *England*, in the counties of Bucks, Chester, Cornwall, Cumberland, Derby, Devon, Dorset, Essex, Gloucester, Hants (the mainland and the Isle of Wight), Hereford, Hertford, Kent, Lancaster, Middlesex, Monmouth, Norfolk, Northampton, Northumberland, Nottingham, Oxford, Salop, Somerset, Stafford, Suffolk, Surrey, Sussex, Warwick, Westmoreland, Wilts, Worcester, and York. In *Wales*, in the counties of Anglesea, Brecknock, Caermarthen, Caernarvon, Cardigan, Denbigh, Glamorgan, Merioneth, Montgomery, and Pembroke. In *Scotland*, in the counties of Argyle, Ayr, Berwick, Dumfries, Kirkcudbright, Lanark, Perth, and Renfrew. In *Ireland*, in the counties of Antrim, Clare, Cork, Down, Dublin, Galway, Kerry, Kilkenny, Limerick, Louth, Sligo, Tipperary Waterford, and Wicklow. In Jersey.

XLVIII.—THE TUNBRIDGE FILMY FERN.

Hymenophyllum tunbridgense.

(Plate XV., Fig. 5, page 77.)

LENGTH OF FROND.—One to six inches, the maximum length being exceptional.

GENERAL DESCRIPTION.—*Roots* very fine, fibrous, wiry, and abundant. *Rootstock*, a very slender, hairlike rhizoma, which branches and creeps extensively, forming oftentimes, with the roots, a dense, matted network, that extends for several yards—the interwoven fibres making a mass that may be stripped off like a thick carpet from the surface of the rock upon which they have spread. *Fronds* evergreen, ovate, and peculiar in conformation. The stipes is brownish-black and hairlike, the rachis continuing it being of similar texture, size, and colour. From each side of the rachis, in alternation, are secondary forked rachides, similar in character to, but somewhat more delicate than, the stipes and primary rachis. The whole of the black, vein-like rachides are margined on either side by semi-pellucid, olive-green, finely-toothed, leaf-like expansions—each side-branch or pinna looking somewhat like the spread fingers of a hand. *Fructification* borne not on the under sides of the leafy parts of the frond, as is the case with the large majority of ferns, but in little cup-shaped indusia, situated upon aborted veins, which branch from the secondary rachides near where these make angles with the main rachis on either side of the latter. The upper margins of the indusia are fringed (see page 18, left-hand figure).

HABITATS.—The damp surfaces of rocks in moist moorland or mountainous country. *Hymenophyllum tunbridgense* is oftentimes found growing in company with mosses either on rocks, on tree-trunks, or on the ground. It ·is also found on boulder rocks in mid-stream, and generally in or near streams, on rock-covered hills or uplands within the influence of the moist emanations from neighbouring streams ; and the hollows, crevices, or sides of waterfalls are favourite habitats, this species often growing almost in darkness in rocky fissures, whose external and frequently internal sides it completely drapes.   A very slight depth of earth suffices for root-room, and oftentimes the carpet of its matted roots and rhizomas appears to cover nothing but the moist surface of bare rocks.

WHERE FOUND.—In *England*, in the counties of Chester, Cornwall, Cumberland, Derby, Devon, Kent, Lancaster, Northumberland, Somerset, Stafford, Sussex, Westmoreland, and York.   The following are the espe-cial localities for this species in the counties named. In the county of Chester : the neighbourhood of Buxton, Croydon Brook, and Macclesfield.   In Cornwall, Rough Tor, near Camelford, and the vicinity of Penryn.   In Devonshire, on Dartmoor, namely, at Becky Fall (near Moreton Hampstead), in Bickleigh Vale, by Shaugh Bridge, on Staple Tor, and on Vixen Tor.   In Kent, vicinity of Tunbridge Wells.   In the county of Lancaster, Cliviger, Conistone, Greenfield, and Rake-Hey Common. In Somersetshire, near Shepton Mallet.   In Sussex, Ardingly, Balcombe, Cockbush (Chichester), Hand-cross (Tilgate Forest), and West Hoathley.   In Yorkshire, the vicinity of Halifax and Esk Dale, neighbourhood of Whitby.   In *Wales*, in the coun-ties of Anglesea, Brecknock, Caernarvon, Glamorgan, and Merioneth ; and in the following localities : in Glamorganshire, Cilhepste, Waterfall, Melincourt Water-fall, and Pont-nedd-Vechan.   In Merionethshire, Cader Idris, Cwm Bychan (in the vicinity of Barmouth), Crafnant (in the neighbourhood of Harlech), Dolgelly, Vale of Festiniog, and Rhaiadr Du (in the neighbour-

hood of Maentwrog). In *Scotland*, in the counties of
Argyle, Dumbarton, Dumfries, Peebles, Renfrew, Ross,
and Stirling. The following are the localities of these
counties :—In Argyleshire, Bullwood, Dunoon, and
Glen Gilp. In Dumbartonshire, shores of Loch
Lomond. In Dumfriesshire, Drumlanrig ; and in
Lanarkshire, banks of River Clyde. It is also found in
the islands of Bute and Mull. In *Ireland*, in the
counties of Clare, Cork, Dublin, Galway, Kerry, and
Wicklow, the subjoined being the localities. In the
county of Clare, Feacle. In Cork, Ballenhassig
Waterfall, Dunbullogue Glen, Glenbower, Glengariff,
Killeagh, and Lota Wood. In the county of Dublin,
in the neighbourhood of the capital. In Galway,
Ballynahinch and Connemara. In Kerry, in Glen
Carnn and the vicinity of Killarney. In the county of
Wicklow, Glencree. *Hymenophyllum tunbridgense* is
found at various elevations, extending to about a
thousand or twelve hundred feet above the sea-level.

XLIX.—THE ONE-SIDED FILMY FERN.

Hymenophyllum unilaterale.

(Plate XV., Fig. 6, page 77.)

LENGTH OF FROND.—One to six inches, the maximum length being exceptional, and the average seldom
exceeding two or three inches.

GENERAL DESCRIPTION.—*Roots* very fine, wiry,
fibrous, and abundant. *Rootstock*, a slender, hairlike,
brownish-black rhizoma, which, like that of *Hymenophyllum tunbridgense*, creeps extensively along the rocks or
shallow soil on which it grows, forming frequently, with
the roots, dense, compact clusters, which are often

intimately mixed with roots of moss, and of its con
gener, the Tunbridge Filmy Fern. *Fronds* evergreen,
elongated, oval in shape ; stipes and rachis brownish-
black ; leafy part olive-green, bipinnate ; pinnæ opposite
or alternate, divided into elongated, narrow pinnules,
which arise from one side—and that the upper—of the
midvein of each pinna. The texture of the fronds is of
the same semi-pellucid nature as that of the fronds of
Hymenophyllum tunbridgense, and they have the appearance as of winged leafy margins to a series of forked
veins,—the distinction between the two species consisting in the fact that the pinnules of *Hymenophyllum
unilaterale*, besides being wider apart from each other,
are produced upon one side only of the pinnæ, and not
on both as is the case in *Hymenophyllum tunbridgense.*
Fructification produced in urn-shaped indusia similar to
those of the Tunbridge Filmy Fern, but entire, instead
of being fringed upon their upper margins—the indusia
being situated upon aborted veins that branch from the
pinnæ on each side of and near the junction of the latter
with the main rachis (see page 18, right-hand figure).

HABITATS.—Exactly similar to those indicated in the
case of *Hymenophyllum tunbridgense,*—namely, damp,
shady rocks, tree-trunks, and the ground, oftentimes
keeping company with that species, and with moss, the
roots and rhizomas interlacing with the mossy roots and
stems.

WHERE FOUND.—In *England*, in the counties of
Cornwall, Cumberland, Devon, Kent, Lancaster, Salop,
Stafford, Westmoreland, and York. The following are the
localities of these counties :—In Cornwall, the vicinity
of Bodmin, Rough Tor (near Camelford), Granite Tor,
and Carn Brea (near Redruth). In the county of Cumberland, Borrowdale, Bow Fell, Scale Force (near Butter-
mere), Dalegarth, Ennerdale, Gatesgarth Dale, High
Still, Honister Crag, Keswick, and Lodore Fall. In
Devonshire, Bickleigh Wood, Moreton Hampstead,

Shaugh Bridge, West Lyn, Wistman's Wood; and on
the following tors : Great Mist, Longaford, Sheep's,
White, and Vixen tors.   In Lancashire, Thevilly (near
• Burnley), neighbourhood of Bury, in caves near Green-
field, and near Lancaster.   In the county of Northumber-
land, Jurionside.   In the county of Salop, Treflack
Wood (Oswestry).   In Staffordshire, Gradbitch (near
Flash).   In Westmoreland, Ambleside, Langdale Pikes,
Patterdale, and Stock Gill Force.   In Yorkshire, Lower
Harrogate, Hawl Gill (near Mickleton), and Turner's
Clough (Rushworth).   In *Wales*, in the counties of
Anglesea, Brecknock, Caermarthen, Caernarvon, Cardi-
gan, Glamorgan, Merioneth, and Radnor.   In Caernar-
vonshire, Capel Curig (near Llanberis), Cwm Idwal,
Rhaiadr Mawr, and Rhaiadr-y-Wenol.   In the county of
Cardigan, Devil's Bridge, Hafod, and Pont Bren.   In
Glamorganshire, Melincourt Waterfall, Scudeinon-Gam.
In the county of Merioneth, Cader Idris, Dolgelly,
Festiniog, Rhaiadr-y-Mawddach (near Llanelltyd), and
Rhaiadr-Du (near Maentwrog).   In *Scotland*, in the
counties of Aberdeen, Argyle, Ayr, Clackmannan,
Dumbarton, Dumfries, Fife, Forfar, Inverness, Kinross,
Kirkcudbright, Orkney (including Shetland), Peebles,
Perth, Renfrew, Ross, Stirling, and Sutherland, the
localities in these counties being : In Argyleshire,
Crinan, Dunoon, Glen Finnart, Glen Gilp, and Glen
Moray.   In Ayrshire, Dalmellington and Glen Ness.
In the county of Clackmannan, Castle Campbell and
Dollar.   In Dumbartonshire, Bowling Hills and· shores
of Loch Lomond.   In the county of Dumfries, Delvine
Pass, Grey Mare's-tail, Girpel Lane, Kirkpatrick-juxta,
Moffat Dale, and Nithside.   In Forfarshire, Reeky
Linn.   In the county of Perth, Ben Lawers, Finlarig
Burn (near Killin), Glen Queich, the Ochils, Pass of
Leny, shores of Lock Katrine, and the Trosachs.   In
the county of Renfrew, Gourock.   In the islands of
Arran, Harris, Islay, and Mull.   In *Ireland*, in the

counties of Antrim, Cork, Donegal, Dublin, Galway, Kerry, Londonderry, Mayo, Tipperary, and Wicklow, the following being the localities of these counties : In the county of Antrim, Colin Glen (Belfast), Glenarve River (Cushendall). In Cork, Morgan's Glen (Clonmel), and near Youghal. In the county of Donegal, the Ennishowan Mountains. In Galway, Connemara and Oughterard. In Kerry, Killarney and the mountains of the county. In Mayo, the mountains of the county. In the county of Wicklow, Glendalough, Hermitage Glen, and Powerscourt Waterfall. *Hymenophyllum unilaterale* is found growing at various heights extending to two thousand eight hundred feet above sea-level.

I.—FERNS ROUND LONDON.

THE number of those in the Metropolis who are lovers and growers of ferns is enormously large, and has certainly largely increased within the last few years. A walk through almost any street will prove the accuracy of this statement, by showing how many ferns are now grown in windows alone. These beautiful, flowerless plants have, in such positions, to a large extent, taken the place which used to be occupied by flowers or other ornaments. Similar evidence of the direction of the popular taste is afforded by the appearance of front suburban gardens.

"Where to find ferns round London?" is, therefore, a question that is being continually asked, and, though the present chapter will not profess to return an exhaustive answer to the inquiry, it will give information which, it is hoped, will be useful and valuable to a large number of persons.

The rapid changes that, by the continual development of London, are made upon the country around it, render it difficult to accurately define the locality of fern habitats; and any attempt to name *particular spots* where ferns are to be found would involve the risk of constant disappointment. Particular habitats may have been stripped, and yet the same ferns may be found in the vicinity of the old habitats. The plants may, so to speak, have been driven further afield ; but the places that used to know them are almost certain to furnish a more or less reliable key to their actual "whereabouts": that is to say, that the old habitat will at least provide or suggest a good starting-point from which to search for the new one.

Amongst the authorities consulted for the purposes of this chapter are the "Flora of Middlesex," by Messrs. Trimen and Dyer, the floras of other metropolitan counties, and Dr. E. de Crespigny's "New London Flora."

The localities are set out in alphabetical order, and the name of each *district* is given in preference to indicating the exact wood, lane, common, or down where the habitat is to be looked for. To direct thousands of , persons, for instance, to the particular part of a wood, lane, or common where certain species of ferns are to be found, would be to secure the speedy extermination of the plants ; and such easy acquisition would take away half of the pleasure of fern-hunting.

With regard especially to the following lists of ferns round London, the Author will be glad at all times to receive from correspondents information supplementary to that contained in this chapter ; and, whenever possible, specimen fronds of ferns found in localities not mentioned here, or not included under the names of the districts which have been mentioned.

ABBEY WOOD. *Lastrea dilatata* (Broad Buckler Fern), *Lastrea filix-mas* (Male Fern).

ACTON.—*Asplenium ruta-muraria* (Rue-leaved Spleenwort), *Ophioglossum vulgatum* (Adders-tongue).

ADDINGTON HILLS. *Blechnum spicant* (Hard Fern), *Botrychium lunaria* (Moonwort).

ALBURY. *Asplenium adiantum-nigrum* (Black Maidenhair Spleenwort), *Asplenium ruta-muraria* (Rue-leaved Spleenwort), *Botrychium lunaria* (Moonwort), *Cystopteris fragilis* (Brittle Bladder Fern), *Lastrea filix-mas* (Male Fern), *Lastrea spinulosa* (Prickly-toothed Buckler Fern), *Ophioglossum vulgatum* (Adders-tongue).

ALDENHAM. *Athyrium filix-fœmina* (Lady Fern), *Lastrea dilatata* (Broad Buckler Fern).

ARDINGLY. *Hymenophyllum tunbridgense* (Tunbridge Filmy Fern), *Lastrea recurva* (Hay-scented Buckler Fern).

ASHTEAD. *Lastrea spinulosa* (Prickly-toothed Buckler Fern).

BADDOW (LITTLE). *Lastrea montana* (Mountain Buckler Fern), *Lastrea thelypteris* (Marsh Buckler Fern), *Osmunda regalis* (Royal Fern).

BAGSHOT. *Athyrium filix-fœmina* (Lady Fern), *Blechnum spicant* (Hard Fern), *Lastrea dilatata* (Broad Buckler Fern), *Lastrea filix-mas* (Male Fern), *Lastrea spinulosa* (Prickly-toothed Buckler Fern), *Osmunda regalis* (Royal Fern), *Polypodium vulgare* (Common Polypody), *Pteris aquilina* (Bracken).

BALCOMBE. *Hymenophyllum tunbridgense* (Tunbridge Filmy Fern), *Lastrea recurva* (Hay-scented Buckler Fern).

BANBURY. *Ophioglossum vulgatum* (Adders-tongue).

BARKING. *Polystichum angulare* (Soft Prickly Shield Fern).

BARNES. *Pteris aquilina* (Bracken).

BARNET. *Lastrea filix-mas* (Male Fern).

BAYFORD. *Asplenium trichomanes* (Common Maidenhair Spleenwort), *Lastrea montana* (Mountain Buckler Fern), *Ophioglossum vulgatum* (Adders-tongue).

BEDDINGTON. *Ophioglossum vulgatum* (Adders-tongue).

BERKHAMPSTEAD (GREAT). *Athyrium filix-fœmina* (Lady Fern), *Lastrea montana* (Mountain Buckler Fern), *Polystichum aculeatum* (Hard Prickly Shield Fern).

BERKHAMPSTEAD (LITTLE). *Asplenium adiantum-nigrum* (Black Maidenhair Spleenwort), *Asplenium trichomanes* (Common Maidenhair Spleenwort), *Athyrium filix-fœmina* (Lady Fern), *Polystichum aculeatum* (Hard Prickly Shield Fern).

BETCHWORTH. *Asplenium ceterach* (Scaly Spleenwort), *Ophioglossum vulgatum* (Adders-tongue).

BEXLEY. *Lastrea montana* (Mountain Buckler Fern), *Lastrea thelypteris* (Marsh Buckler Fern).

BLACK NOTLEY. *Polystichum aculeatum* (Hard Prickly Shield Fern).

BLETCHINGLEY. *Asplenium ruta-muraria* (Rue-leaved Spleenwort), *Polystichum aculeatum* (Hard Prickly Shield Fern).

L

BRASTED.   *Lastrea montana* (Mountain Buckler Fern).

BRENTFORD.   *Ophioglossum vulgatum* (Adders-tongue), *Polystichum angulare* (Soft Prickly Shield Fern).

BRENTWOOD.   *Athyrium filix-fœmina* (Lady Fern), *Blechnum spicant* (Hard Fern), *Lastrea dilatata* (Broad Buckler Fern), *Lastrea montana* (Mountain Buckler Fern), *Lastrea spinulosa* (Prickly-toothed Buckler Fern), *Ophioglossum vulgatum* (Adders-tongue), *Osmunda regalis* (Royal Fern), *Polystichum aculeatum* (Hard Prickly Shield Fern).

BRICKENDON.   *Asplenium adiantum-nigrum* (Black Maidenhair Spleenwort), *Ophioglossum vulgatum* (Adders-tongue), *Polystichum angulare* (Soft Prickly Shield Fern).

BROCKHAM.   *Ophioglossum vulgatum* (Adders-tongue).

BROXBOURNE.   *Blechnum spicant* (Hard Fern), *Lastrea filix-mas* (Male Fern), *Lastrea dilatata* (Broad Buckler Fern), *Lastrea montana* (Mountain Buckler Fern), *Lastrea spinulosa* (Prickly-toothed Buckler Fern).

BURNHAM BEECHES.   *Asplenium adiantum-nigrum* (Black Maidenhair Spleenwort), *Asplenium trichomanes* (Common Maidenhair Spleenwort), *Athyrium filix-fœmina* (Lady Fern), *Blechnum spicant* (Hard Fern), *Lastrea filix-mas* (Male Fern), *Polystichum aculeatum* (Hard Prickly Shield Fern), *Polystichum angulare* (Soft Prickly Shield Fern).

BURSTOW.   *Asplenium trichomanes* (Common Maidenhair Spleenwort).

CANTERBURY.   *Lastrea spinulosa* (Prickly-toothed Buckler Fern), *Ophioglossum vulgatum* (Adders-tongue).

CHALFONT.   *Polystichum aculeatum* (Hard Prickly Shield Fern).

CHERTSEY.   *Lastrea dilatata* (Broad Buckler Fern).

CHESHUNT.   *Polystichum aculeatum* (Hard Prickly Shield Fern).

CHIDDINGLY.   *Hymenophyllum tunbridgense* (Tunbridge Filmy Fern).

CHIGWELL.   *Lastrea thelypteris* (Marsh Buckler Fern), *Polystichum aculeatum* (Hard Prickly Shield Fern).

CHINGFORD.   *Polystichum aculeatum* (Hard Prickly Shield Fern), *Pteris aquilina* (Bracken).

CHIPPING NORTON.   *Lastrea dilatata* (Broad Buckler Fern).

CHISLEHURST.   *Botrychium lunaria* (Moonwort), *Lastrea spinulosa* (Prickly-toothed Buckler Fern), *Pteris aquilina* (Bracken), *Scolopendrium vulgare* (Hartstongue).

CHOBHAM.   *Blechnum spicant* (Hard Fern), *Lastrea dilatata* (Broad Buckler Fern), *Lastrea filix-mas* (Male Fern), *Lastrea montana* (Mountain Buckler Fern), *Lastrea spinulosa* (Prickly-toothed Buckler Fern), *Osmunda regalis* (Royal Fern).

COBHAM.   *Lastrea filix-mas* (Male Fern), *Lastrea montana* (Mountain Buckler Fern), *Lastrea spinulosa* (Prickly-toothed

Buckler Fern), *Ophioglossum vulgatum* (Adders-tongue), *Polypodium vulgare* (Common Polypody), *Polystichum aculeatum* (Hard Prickly Shield Fern).

COGGLESHALL. *Lastrea spinulosa* (Prickly-toothed Buckler Fern).

COLDHARBOUR. *Athyrium filix-fœmina* (Lady Fern), *Lastrea dilatata* (Broad Buckler Fern), *Lastrea filix-mas* (Male Fern), *Lastrea spinulosa* (Prickly-toothed Buckler Fern), *Osmunda regalis* (Royal Fern), *Polypodium vulgare* (Common Polypody), *Polystichum aculeatum* (Hard Prickly Shield Fern).

COLNEY HEATH. *Athyrium filix-fœmina* (Lady Fern), *Lastrea filix-mas* (Male Fern), *Polystichum aculeatum* (Hard Prickly Shield Fern), *Polystichum angulare* (Soft Prickly Shield Fern).

COULSDON. *Botrychium lunaria* (Moonwort), *Ophioglossum vulgatum* (Adders-tongue), *Pteris aquilina* (Bracken).

COWLEY. *Asplenium ceterach* (Scaly Spleenwort).

CRAY (NORTH). *Lastrea thelypteris* (Marsh Buckler Fern).

CRAY (ST. MARY). *Polystichum angulare* (Soft Prickly Shield Fern).

CROHAM HURST. *Polypodium vulgare* (Common Polypody).

CROYDON. *Ophioglossum vulgatum* (Adders-tongue).

DANBURY. *Lastrea spinulosa* (Prickly-toothed Buckler Fern).

DARTFORD. *Asplenium adiantum-nigrum* (Black Maidenhair Spleenwort), *Athyrium filix-fœmina* (Lady Fern), *Blechnum spicant* (Hard Fern), *Botrychium lunaria* (Moonwort), *Lastrea filix-mas* (Male Fern), *Ophioglossum vulgatum* (Adders-tongue), *Pteris aquilina* (Bracken).

DORKING. *Botrychium lunaria* (Moonwort), *Lastrea dilatata* (Broad Buckler Fern), *Osmunda regalis* (Royal Fern), *Polypodium vulgare* (Common Polypody), *Polystichum aculeatum* (Hard Prickly Shield Fern).

EARLSWOOD. *Polystichum aculeatum* (Hard Prickly Shield Fern).

ELSTEAD. *Blechnum spicant* (Hard Fern), *Lastrea filix-mas* (Male Fern), *Lastrea spinulosa* (Prickly-toothed Buckler Fern).

ELSTREE. *Ophioglossum vulgatum* (Adders-tongue).

EPPING. *Asplenium ruta-muraria* (Rue-leaved Spleenwort), *Lastrea dilatata* (Broad Buckler Fern), *Lastrea filix-mas* (Male Fern), *Lastrea spinulosa* (Prickly-toothed Buckler Fern), *Lastrea thelypteris* (Marsh Buckler Fern), *Osmunda regalis* (Royal Fern), *Polystichum angulare* (Soft Prickly Shield Fern).

EPPING FOREST. *Asplenium ruta-muraria* (Rue-leaved Spleenwort), *Asplenium ceterach* (Scaly Spleenwort), *Blechnum spicant* (Hard Fern), *Lastrea dilatata* (Broad Buckler Fern), *Lastrea montana* (Mountain Buckler Fern), *Lastrea spinulosa* (Prickly-toothed Fern), *Lastrea thelypteris* (Marsh Buckler Fern), *Ophioglossum vulgatum* (Adders-tongue), *Polypodium vulgare* (Common

Polypody), *Polystichum aculeatum* (Hard Prickly Shield Fern), *Polystichum angulare* (Soft Prickly Shield Fern), *Pteris aquilina* (Bracken), *Scolopendrium vulgare* (Hartstongue).

EPSOM. *Pteris aquilina* (Bracken).

ESHER. *Blechnum spicant* (Hard Fern), *Lastrea dilatata* (Broad Buckler Fern), *Lastrea spinulosa* (Prickly-toothed Buckler Fern), *Osmunda regalis* (Royal Fern), *Pteris aquilina* (Bracken).

ESSENDON. *Asplenium adiantum-nigrum* (Black Maidenhair Spleenwort), *Ophioglossum vulgatum* (Adders-tongue), *Polystichum aculeatum* (Hard Prickly Shield Fern).

EWHURST. *Osmunda regalis* (Royal Fern).

FARLEIGH (WEST). *Ophioglossum vulgatum* (Adders-tongue).

FARNHAM. *Asplenium ceterach* (Scaly Spleenwort), *Botrychium lunaria* (Moonwort), *Osmunda regalis* (Royal Fern).

FOOT'S CRAY. *Asplenium trichomanes* (Common Maidenhair Spleenwort), *Botrychium lunaria* (Moonwort).

FRENSHAM. *Polypodium vulgare* (Common Polypody).

FRIMLEY. *Osmunda regalis* (Royal Fern).

FULMER. *Lastrea spinulosa* (Prickly-toothed Buckler Fern), *Polystichum aculeatum* (Hard Prickly Shield Fern).

GERRARD'S CROSS. *Pteris aquilina* (Bracken).

GODALMING. *Asplenium ceterach* (Scaly Spleenwort), *Botrychium lunaria* (Moonwort), *Lastrea thelypteris* (Marsh Buckler Fern), *Osmunda regalis* (Royal Fern), *Polypodium vulgare* (Common Polypody). Also (at Hascombe) *Lastrea filix-mas, Lastrea spinulosa,* and *Polystichum aculeatum* (W. A. Pearce).

GODSTONE. *Asplenium ruta-muraria* (Rue-leaved Spleenwort), *Osmunda regalis* (Royal Fern).

GOMSHALL. *Asplenium trichomanes* (Common Maidenhair Spleenwort), *Athyrium filix-fœmina* (Lady Fern), *Lastrea spinulosa,* (Prickly-toothed Buckler Fern).

GRAVESEND. *Asplenium ceterach* (Scaly Spleenwort).

GREENFORD. *Ophioglossum vulgatum* (Adders-tongue).

GREENHITHE. *Ophioglossum vulgatum* (Adders-tongue).

GUILDFORD. *Asplenium adiantum-nigrum* (Black Maidenhair Spleenwort), *Asplenium ruta-muraria* (Rue-leaved Spleenwort), *Asplenium trichomanes* (Common Maidenhair Spleenwort), *Athyrium filix-fœmina* (Lady Fern), *Cystopteris fragilis* (Brittle Bladder Fern), *Lastrea dilatata* (Broad Buckler Fern) *Lastrea filix-mas* (Male Fern), *Lastrea montana* (Mountain Buckler Fern), *Lastrea spinulosa* (Prickly-toothed Buckler Fern), *Ophioglossum vulgatum* (Adders-tongue), *Polystichum angulare* (Soft Prickly Shield Fern), *Pteris aquilina* (Bracken).

HACKNEY MARSHES. *Ophioglossum vulgatum* (Adders-tongue).

HAINAULT FOREST. *Lastrea thelypteris* (Marsh Buckler Fern), *Polystichum aculeatum* (Hard Prickly Shield Fern).

HAMPSTEAD HEATH. *Asplenium ruta-muraria* (Rue-leaved Spleenwort), *Pteris aquilina* (Bracken).

HANDCROSS. *Hymenophyllum tunbridgense* (Tunbridge Filmy Fern).

HAREFIELD. *Asplenium adiantum-nigrum* (Black Maidenhair Spleenwort), *Asplenium trichomanes* (Common Maidenhair Spleenwort), *Lastrea cristata* (Crested Buckler Fern), *Lastrea dilatata* (Broad Buckler Fern), *Ophioglossum vulgatum* (Adders-tongue), *Polystichum aculeatum* (Hard Prickly Shield Fern).

HARROW WEALD. *Asplenium adiantum-nigrum* (Black Maidenhair Spleenwort), *Athyrium filix-fœmina* (Lady Fern), *Blechnum spicant* (Hard Fern), *Lastrea dilatata* (Broad Buckler Fern), *Lastrea montana* (Mountain Buckler Fern), *Polystichum angulare* (Soft Prickly Shield Fern), *Pteris aquilina* (Bracken).

HARTWELL. *Lastrea montana* (Mountain Buckler Fern).

HASLEMERE. *Asplenium adiantum-nigrum* (Black Maidenhair Spleenwort), *Asplenium ceterach* (Scaly Spleenwort), *Asplenium trichomanes* (Common Maidenhair Spleenwort).

HATFIELD. *Asplenium adiantum-nigrum* (Black Maidenhair Spleenwort), *Asplenium trichomanes* (Common Maidenhair Spleenwort), *Athyrium filix-fœmina* (Lady Fern), *Blechnum spicant* (Hard Fern), *Lastrea spinulosa* (Prickly-toothed Buckler Fern), *Polystichum angulare* (Soft Prickly Shield Fern).

HAYES. *Pteris aquilina* (Bracken).

HENDON. *Polystichum aculeatum* (Hard Prickly Shield Fern).

HERTFORD. *Asplenium adiantum-nigrum* (Black Maidenhair Spleenwort), *Asplenium trichomanes* (Common Maidenhair Spleenwort), *Athyrium filix-fœmina* (Lady Fern), *Blechnum spicant* (Hard Fern), *Lastrea dilatata* (Broad Buckler Fern), *Lastrea filix-mas* (Male Fern), *Lastrea spinulosa* (Prickly-toothed Buckler Fern), *Polystichum angulare* (Soft Prickly Shield Fern).

HERTINGFORDBURY. *Polystichum angulare* (Soft Prickly Shield Fern).

HIGH BEECH. *Lastrea montana* (Mountain Buckler Fern), *Pteris aquilina* (Bracken).

HIGHGATE. *Asplenium ruta-muraria* (Rue-leaved Spleenwort).

HITCHIN. *Lastrea dilatata* (Broad Buckler Fern), *Ophioglossum vulgatum* (Adders-tongue), *Polystichum aculeatum* (Hard Prickly Shield Fern).

HOLMWOOD. *Blechnum spicant* (Hard Fern), *Lastrea montana* (Mountain Buckler Fern), *Osmunda regalis* (Royal Fern), *Pteris aquilina* (Bracken).

HORSELL. *Lastrea montana* (Mountain Buckler Fern).

ISLEWORTH. *Ophioglossum vulgatum* (Adders-tongue).

KELVEDON. *Asplenium trichomanes* (Common Maidenhair Spleenwort).

KESTON. *Asplenium adiantum-nigrum* (Black Maidenhair Spleenwort), *Blechnum spicant* (Hard Fern), *Lastrea filix-mas* (Male Fern), *Lastrea thelypteris* (Marsh Buckler Fern), *Polypodium vulgare* (Common Polypody), *Pteris aquilina* (Bracken).

LEATHERHEAD. *Polystichum aculeatum* (Hard Prickly Shield Fern).

LEIGH. *Polypodium vulgare* (Common Polypody), *Scolopendrium vulgare* (Hartstongue).

LEITH HILL. *Blechnum spicant* (Hard Fern), *Botrychium lunaria* (Moonwort), *Lastrea dilatata* (Broad Buckler Fern), *Lastrea filix-mas* (Male Fern), *Lastrea montana* (Mountain Buckler Fern), *Lastrea spinulosa* (Prickly-toothed Buckler Fern), *Lastrea thelypteris* (Marsh Buckler Fern), *Osmunda regalis* (Royal Fern), *Polystichum aculeatum* (Hard Prickly Shield Fern).

LEYTON. *Asplenium trichomanes* (Common Maidenhair Spleenwort).

LEYTONSTONE. *Asplenium ruta-muraria* (Rue-leaved Spleenwort).

LOUGHTON. *Asplenium adiantum-nigrum* (Black Maidenhair Spleenwort), *Lastrea thelypteris* (F. J. Lewis), *Pteris aquilina* (Bracken).

MAIDSTONE. *Asplenium ceterach* (Scaly Spleenwort), *Lastrea filix-mas* (Male Fern).

MAYFORD. *Athyrium filix-fœmina* (Lady Fern), *Lastrea filix-mas* (Male Fern), *Polystichum aculeatum* (Hard Prickly Shield Fern), *Polystichum angulare* (Soft Prickly Shield Fern).

MERSTHAM. *Ophioglossum vulgatum* (Adders-tongue).

MICKLEHAM. *Asplenium ceterach* (Scaly Spleenwort).

MIMMS (North). *Blechnum spicant* (Hard Fern), *Lastrea spinulosa* (Prickly-toothed Buckler Fern), *Polystichum angulare* (Soft Prickly Shield Fern).

MUNCOMBE. *Polystichum aculeatum* (Hard Prickly Shield Fern).

NEWLAND. *Lastrea dilatata* (Broad Buckler Fern).

NORTHAW. *Lastrea montana* (Mountain Buckler Fern).

NUTFIELD. *Asplenium trichomanes* (Common Maidenhair Spleenwort), *Polystichum aculeatum* (Hard Prickly Shield Fern).

ONGAR. *Asplenium adiantum-nigrum* (Black Maidenhair Spleenwort), *Asplenium trichomanes* (Common Maidenhair Spleenwort), *Athyrium filix-fœmina* (Lady Fern), *Ophioglossum vulgatum* (Adders-tongue), *Polystichum aculeatum* (Hard Prickly Shield Fern), *Scolopendrium vulgare* (Hartstongue).

OXHEY. *Athyrium filix-fœmina* (Lady Fern), *Lastrea dilatata* (Broad Buckler Fern), *Ophioglossum vulgatum* (Adders-tongue).

PERIVALE. *Ophioglossum vulgatum* (Adders-tongue).

PINNER. *Asplenium adiantum-nigrum* (Black Maidenhair Spleenwort), *Lastrea dilatata* (Broad Buckler Fern), *Lastrea filix-*

mas (Male Fern), *Polystichum aculeatum* (Hard Prickly Shield Fern), *Polystichum angulare* (Soft Prickly Shield Fern).

PIRBRIGHT. *Blechnum spicant* (Hard Fern), *Lastrea filix-mas* (Male Fern), *Lastrea thelypteris* (Marsh Buckler Fern), *Osmunda regalis* (Royal Fern), *Pteris aquilina* (Bracken).

PUTNEY. *Pteris aquilina* (Bracken).

PUTTENHAM. *Blechnum spicant* (Hard Fern), *Botrychium lunaria* (Moonwort).

RAINHAM. *Asplenium trichomanes* (Common Maidenhair Spleenwort).

REDHILL. *Scolopendrium vulgare* (Hartstongue).

REIGATE. *Asplenium trichomanes* (Common Maidenhair Spleenwort), *Athyrium filix-fœmina* (Lady Fern), *Blechnum spicant* (Hard Fern), *Botrychium lunaria* (Moonwort), *Lastrea dilatata* (Broad Buckler Fern), *Lastrea filix-mas* (Male Fern), *Lastrea spinulosa* (Prickly-toothed Buckler Fern), *Ophioglossum vulgatum* (Adderstongue), *Osmunda regalis* (Royal Fern), *Polystichum aculeatum* (Hard Prickly Shield Fern), *Polystichum angulare* (Soft Prickly Shield Fern), *Pteris aquilina* (Bracken).

RICKMANSWORTH. *Asplenium adiantum-nigrum* (Black Maidenhair Spleenwort), *Asplenium trichomanes* (Common Maidenhair Spleenwort).

RIVERHEAD. *Asplenium ceterach* (Scaly Spleenwort).

RUSTHALL. *Lastrea montana* (Mountain Buckler Fern).

ST. ALBANS. *Asplenium adiantum-nigrum* (Black Maidenhair Spleenwort), *Polystichum aculeatum* (Hard Prickly Shield Fern).

SEVENOAKS. *Lastrea filix-mas* (Male Fern), *Pteris aquilina* (Bracken), *Scolopendrium vulgare* (Hartstongue).

SHACKLEFORD. *Botrychium lunaria* (Moonwort).

SHALFORD. *Asplenium ruta-muraria* (Rue-leaved Spleenwort), *Lastrea dilatata* (Broad Buckler Fern).

SHIERE. *Asplenium adiantum-nigrum* (Black Maidenhair Spleenwort), *Asplenium ruta-muraria* (Rue-leaved Spleenwort), *Asplenium trichomanes* (Common Maidenhair Spleenwort) *Botrychium lunaria* (Moonwort), *Lastrea dilatata* (Broad Buckler Fern), *Lastrea montana* (Mountain Buckler Fern), *Lastrea spinulosa* (Prickly-toothed Buckler Fern), *Osmunda regalis* (Royal Fern).

SHIRLEY. *Athyrium filix-fœmina* (Lady Fern), *Blechnum spicant* (Hard Fern), *Botrychium lunaria* (Moonwort).

SNARESBROOK. *Athyrium filix-fœmina* Lady Fern).

SOUTHBOROUGH. *Blechnum spicant* (Hard Fern), *Pteris aquilina* (Bracken).

SPRINGFIELD. *Polystichum angulare.* (Soft Prickly Shield Fern).

STANMORE. *Lastrea dilatata* (Broad Buckler Fern), *Pteris aquilina* (Bracken).

STURRY. *Lastrea filix-fœmina* (Lady Fern).
SUNNINGHILL. *Lastrea thelypteris* (Marsh Buckler Fern).
TEDDINGTON. *Asplenium adiantum-nigrum* (Black Maiden-
hair Spleenwort), *Asplenium ruta-muraria* (Rue-leaved Spleen-
wort).
TILGATE FOREST. *Asplenium adiantum-nigrum* (Black
Maidenhair Spleenwort), *Athyrium filix-fœmina* (Lady Fern),
Blechnum spicant (Hard Fern), *Hymenophyllum tunbridgense*
(Tunbridge Filmy Fern), *Lastrea dilatata* (Broad Buckler Fern),
Lastrea filix-mas (Male Fern), *Lastrea montana* (Mountain Buckler
Fern), *Lastrea spinulosa* (Prickly-toothed Buckler Fern), *Poly-
podium dryopteris* (Three-branched Polypody), *Polypodium
phegopteris* (Mountain Polypody), *Scolopendrium vulgare* (Harts-
tongue).
TIPTREE. *Lastrea spinulosa* (Prickly-toothed Buckler Fern).
TOTTERIDGE. *Polystichum aculeatum* (Hard Prickly Shield
Fern), *Polystichum angulare* (Soft Prickly Shield Fern).
TOWN MALLING. *Asplenium ruta-muraria* (Rue-leaved Spleen-
wort).
TRING. *Athyrium filix-fœmina* (Lady Fern), *Blechnum spicant*
(Hard Fern), *Lastrea filix-mas* (Male Fern), *Lastrea montana*
(Mountain Buckler Fern).
TUNBRIDGE WELLS. *Asplenium lanceolatum* (Lanceolate
Spleenwort), *Asplenium trichomanes* (Common Maidenhair Spleen-
wort), *Athyrium filix-fœmina* (Lady Fern), *Blechnum spicant*
(Hard Fern), *Cystopteris fragilis* (Brittle Bladder Fern), *Hymeno-
phyllum tunbridgense* (Tunbridge Filmy Fern), *Lastrea dilatata*
(Broad Buckler Fern), *Lastrea filix-mas* (Male Fern), *Lastrea
montana* (Mountain Buckler Fern), *Lastrea recurva* (Hay-scented
Buckler Fern), *Lastrea spinulosa* (Prickly-toothed Buckler Fern),
Lastrea thelypteris (Marsh Buckler Fern), *Osmunda regalis* (Royal
Fern), *Polypodium vulgare* (Common Polypody), *Pteris aquilina*
(Bracken), *Scolopendrium vulgare* (Hartstongue).
VIRGINIA WATER. *Athyrium filix-fœmina* (Lady Fern), *Lastrea
dilatata* (Broad Buckler Fern), *Lastrea filix-mas* (Male Fern),
Lastrea spinulosa (Prickly-toothed Buckler Fern).
WALTHAMSTOW. *Asplenium trichomanes* (Common Maidenhair
Spleenwort), *Pteris aquilina* (Bracken).
WANDSWORTH. *Pteris aquilina* (Bracken).
WARLEY. *Asplenium adiantum-nigrum* (Black Maidenhair
Spleenwort), *Asplenium trichomanes* (Common Maidenhair Spleen-
wort), *Athyrium filix-fœmina* (Lady Fern), *Blechnum spicant*
(Hard Fern), *Lastrea dilatata* (Broad Buckler Fern), *Lastrea
montana* (Mountain Buckler Fern), *Lastrea spinulosa* (Prickly-
toothed Buckler Fern), *Ophioglossum vulgatum* (Adders-tongue),
Osmunda regalis (Royal Fern), *Polystichum aculeatum* (Hard

Prickly Shield Fern), *Polystichum angulare* (Soft Prickly Shield Fern).

WATFORD. *Asplenium . adiantum-nigrum* (Black Maidenhair Spleenwort), *Ophioglossum vulgatum* (Adders-tongue), *Polystichum angulare* (Soft Prickly Shield Fern).

WELHAM. *Asplenium adiantum-nigrum* (Black Maidenhair Spleenwort), *Asplenium trichomanes* (Common Maidenhair Spleenwort), *Polystichum aculeatum* (Hard Prickly Shield Fern).

WENDLESHAM. *Polypodium vulgare* (Common Polypody).

WEST HOATHLEY. *Lastrea recurva* (Hay-scented Buckler Fern).

WEYBRIDGE. *Pteris aquilina* (Bracken).

WHETTON. *Lastrea dilatata* (Broad Buckler Fern).

WIMBLEDON. *Polystichum aculeatum* (Hard Prickly Shield Fern), *Pteris aquilina* (Bracken).

WINCHMORE HILL. *Athyrium filix-fœmina* (Lady Fern), *Lastrea dilatata* (Broad Buckler Fern), *Lastrea spinulosa* (Prickly-toothed Buckler Fern), *Polypodium vulgare* (Common Polypody).

WINDSOR. *Athyrium filix-fœmina* (Lady Fern), *Lastrea thelypteris* (Marsh Buckler Fern).

WITLEY. *Blechnum spicant* (Hard Fern), *Lastrea montana* (Mountain Buckler Fern), *Lastrea spinulosa* (Prickly-toothed Buckler Fern), *Osmunda regalis* (Royal Fern).

WOKING. *Blechnum spicant* (Hard Fern), *Lastrea dilatata* (Broad Buckler Fern), *Lastrea spinulosa* (Prickly-toothed Buckler Fern), *Pteris aquilina* (Bracken), *Scolopendrium vulgare* (Harts-tongue).

WONHAM. *Lastrea dilatata* (Broad Buckler Fern).

WOODFORD. *Asplenium ruta-muraria* (Rue-leaved Spleenwort), *Asplenium trichomanes* (Common Maidenhair Spleenwort).

WORMLEY. *Lastrea filix-mas* (Male Fern), *Lastrea montana* (Mountain Buckler Fern).

———•◦•———

INDEX OF LOCALITIES

Referred to between pages 1 and 137.

———◆◇◆———

THE END.

WYMAN AND SONS, PRINTERS, GREAT QUEEN STREET, W C.